AF772774

Airships *R-100* and *R-101*

The Success of the *R-100*'s Trip to Canada and the Tragedy of the *R-101* in France

Michel Pratt

Photo from Rosario Audette. Gilbert Audette Collection.

Michel Pratt, éditeur

You can get this book from :
Michel Pratt
michelpratt.quebec
prattmichel@icloud.com

Pratt, Michel, 1949-
Title: *Airships R-100 and R-101*
Other title: The Success of the *R-100*'s Trip to Canada and the Tragedy of the *R-101* in France
1.- *R-100* and *R-101* (Airships). 2.- Airships-Canada-Québec- Longueuil-Saint-Hubert- History. 3.- Airships-Great Britain-History. 4.- Aeronautics Canada-Québec-History 20th century.

Michel Pratt,éditeur

Translation: Kader Barry
Graphical conception: Michel Pratt and Sébastien Robert
Revision: Raymond Ostiguy and Kevin Henley with the collaboration of Richard Smith.
Administration: Ginette Guilbault
This publication has obtained financial assistance from Emploi-Québec.

Legal Deposit: 2018

Bibliothèque nationale du Québec
National Library of Canada
ISBN: 978-2981695659

Photo on the cover : Airship *R-100* Moored at the Saint-Hubert Tower.
Louis Emard Collection.

ACKNOWLEDGMENTS

We wish like to thank the following persons for the loan of their photos:

Simon Beauregard (Longueuil), Yves Beauregard (Quebec), Pierre Gilbert (Paris, France), Serge Juneau (Saint-Augustin-de-Desmaures), Marie-Paule Shaffer-Levac (Longueuil), Jean Mercure (Longueuil), Wilma Metevier (United States of America), Christian Paquin (Montreal), Raymond Plante (Fenelon Falls, Ontario). We appreciate the generosity of Dan Telfair, Secretary of the Nevil Shute Norway Foundation, for having provided us a photograph of a painting of the likeness of Nevil Shute Norway.

We are particularly grateful to Louis Emard (Saint-Bruno-de-Montarville) and Marc-André Valiquette (Laval), noteworthy collectors of the *R-100*, for providing us withsubstantial materials.

Permission was obtained to publish a few photos from the McCord Museum of Montreal, and from Library and Archives Canada. The Airship Heritage Trust of the United Kingdom allowed us to use images from its collection.

We wish to thank Ginette Guilbault, Administrative Secretary of the Société historiqueet culturelle du Marigot, for her competent management of this project and for her constant encouragement.

CONTENTS

Introduction

The *R-100* moored to its anchor mast, in Saint-Hubert (Longueuil).
Wilma Metevier Collection. (Longueuil).

Only a few events in the history of the Province of Quebec have produced such a reverberating echo as the coming, in 1930, of the *R-100* airship. Not only was the media coverage exceptional, as popular singer Mary Travers (la Bolduc) put it, in one of her songs: “Tout l’monde parle du *R-100*” or “Everyone’s talking about the *R-100*”. From the regional perspective, the flight of the *R-100* was the biggest tourist event in the communities of the Montreal South Shore. From the provincial perspective, it was no less important than Expo 67. In terms of technological challenge, however, the interest was worldwide.

The catastrophe of the *R-101* crash in France had as huge a media fallout in this country as in England. Although relatively short, the life spans of the *R-100* and *R-101* reflected an entire decade of innovation in the history of aviation.

As a majestic spectacle, a technological breakthrough and a media event, the history of the *R-100*, that beautiful silver aerial ship, the Titanic of the sky, is all of that and much more… This book relates the success of the first airship, the *R-100*, and the failure of the second, the *R-101*.

THE EVOLUTION OF THE AIRSHIPS

The brothers Joseph-Michel, and Jacques-Étienne Mongolfier, paper producers from Vidalon-lès-Annonay, in France, were intrigued by how smoke rises. They discovered that heat could make a sheet of paper fly. Their experiments progressed to a point that, on September 19, 1783, they managed, in Versailles, to put aloft a hot air balloon carrying a duck, a rooster and a sheep in a wicker basket. King Louis XVI was present at the event. On November 21, 1783, Pilâtre de Rosier, and the Marquis of Arlande made a 28 minute trip over Paris. The Mongolfier brothers were knighted by Louis XVI as a reward for their invention. The following month, in December, Jacques Charles and Nicolas Robert also made a trip in a balloon covered with fabric and filled with hydrogen instead of hot air. The dirigible, sent up into the air on September 24, 1852, by Jules Henri Giffard (1825-1882), was equipped with a 3 h.p. steam-engine-activated propeller; it measured 43 metres in length. Departing from the hippodrome of Paris, it managed to fly, more or less, over a part of that city at a speed never exceeding 5 km/h.

Another French innovation was born on August 9, 1884, when Charles Renard and Arthur Krebs activated, using an electric motor of 8.5 h.p., the dirigible *La France*, equipped with a gas layer, but still without a rigid structure.

The basic principle for the operation of the airship is quite simple. The machine must be equipped with an envelope containing a product rendering it lighter than air. In the beginning, hydrogen, which only weighs about 1/15th of regular air, was used. Helium, contrary to hydrogen, is not flammable, but it is less effective for lift, and costs more to produce. The form was to remain, the technology to advance. The first triumphs of the rigid structure balloon are well known in the history of aviation. On July 2, 1900, Ferdinand von Zeppelin managed to fly the *LZ-1*, with five persons on board, for more than 15 minutes. That was the beginning of a long adventure. Zeppelin's influence was so great that for a long time people spoke of "zeppelins" rather than dirigibles. The frame of those airships was built from girders of light metal and covered with fabric.

Among the 173 rigid structure airships that flew, there were 130 zeppelins (*LZ-1* to *LZ-130*), 22 *Schütte-Lanz* (*Sl-1* to *Sl-22*), 16 British airships, four American and one French; hence, 152 airships were built by Germans. The development of the zeppelins was Airships, before the end of the First World War, were developing more quickly than the airplanes. Also during that War, a few zeppelins were converted to military purposes.

Mongolfier's balloon.

Giffard's balloon.

Henry Giffard, the inventor.

Certain German models were even used to bomb London. The experiment was however very short-lived as warplanes flew faster than the zeppelins making them a relatively easy target. Since they could remain stationary in the air longer however, zeppelins had a certain advantage over airplanes as observation or reconnaissance tools.

A trip by airship was an often a bold and sometimes a disastrous undertaking. Airships like the Dixmude, the R-34, the USS Shenandoah and the Italia were destroyed in accidents.

The first exploded in middle of the night, on December 21, 1923, above the Mediterranean, close to Sicily, in a sky tormented by winds. In 1919, the second, the *R-34*, had already achieved a feat worthy of mention by flying, in a little more than 108 hours, back and forth, between East Fortune, close to Edinburgh in Scotland, and New York; in January 1921, it crashed in the fog. The USS Shenandoah was the first airship built entirely in the United States; it operated using helium rather than hydrogen. The machine crashed in Ohio on September 3, 1925. The last one, the Italia, crashed at the North Pole, in May 1928. Only the Germans continued to outperform others with their airships. At the same time, airplane speeds increased and flights lasted longer and longer. Mongolfier's balloon.

Graf Zeppelin, the inventor.
Société historique et culturelle du Marigot Collection.

The German airship *Graf Zeppelin*.
Société historique et culturelle du Marigot Collection.

The *Hindenburg* in flames near its mooring mast, in Lakehurst, New Jersey, in 1937.

Construction factory for the *R-100* at Howden.

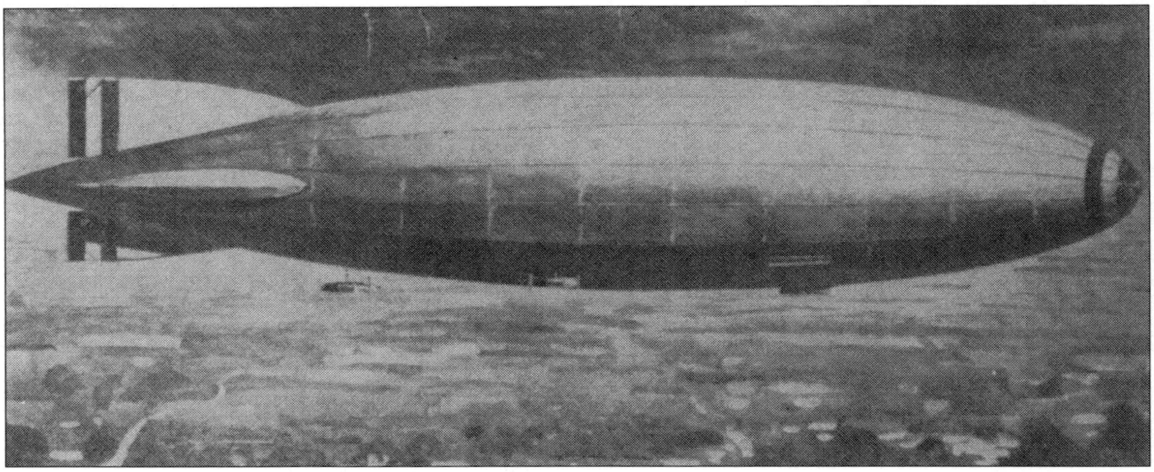

Plan of the model of the *R-100* as originally conceived; it bears a great resemblance to the *R-34*.

Barnes Neville Wallis, designer of the *R-100* airship.

Nevil Shute Norway, «Calculator» of the *R-100*.
Dan Telfair Collection, Secretary

THE CONSTRUCTION OF THE R-100

During the 1920s, despite a few accidents, the use of airships seemed an efficient means of ensuring, at the same time, the transport of airmail, passengers and merchandise. At that time, planes did not offer all these possibilities.

The gigantic dimensions and the roaring of an airship provided, to the countries that owed them, not just a negligible amount of prestige but also a psychological advantage. Nazi Germany, Fascist Italy as well as the USA and France, were building airships. Great Britain, dominant at the time, therefore wanted to equip herself with the biggest of dirigibles.

She planned to stay in touch with her colonies by ensuring transportation by air, whether in her Dominion of Canada or her Empire in India.

The *R-34* crossed the Atlantic during the summer of 1919; unfortunately, it missed its landing, on January 1921, in Howden, Yorkshire. The *R-36* had been built to carry about 50 people; it suffered a lot of damage before being demolished in 1926. The *R-38*, constructed at the same time as the R-34 and destined for the American market, also crashed, on August 23, 1921; 44 of the 49 persons on board, including 17 Americans, perished. The dirigible was cut in half during that accident. After the war, for economic reasons, the British government stopped authorizing the construction of new airships.
In 1922, the Vickers Company proposed to the British Conservative Government the idea of building a few airships capable of making the trip between Great Britain and the British Empire of India. The proposal, presented by Sir Charles Dennistoun Burney, a former Conservative Member from Uxbridge, was accepted during the summer of 1923. However, when the Labour Party took power, in January 1924, Prime Minister James Ramsey McDonald asked for a Committee to revise the brief. It was evident that Great Britain aimed to surpass Germany in the design of airships.
In the spring of 1924, the British Air Ministry, directed by Christopher Birdwood Thomson, entrusted to two different companies the building of two quite similar models: the *R-100* and the *R-101*. The construction of the first one was delegated to a private company, the Airship Guaranty Company. A budget of £300,000 was allocated for its construction. The contract was signed on October 22, 1924. A Crown corporation, the Royal Airship Works, built the *R-101*. In 1925, the Conservative Government of Stanley Baldwin took back the reins of power until 1929.

The Airship Guaranty Company, founded on November 20, 1923, was a subsidiary of the Vickers Company. Its workshops, bought for the sum of £50,000, were located near Howden, in Yorkshire, while those of the Royal Airship Works were in Cardington, in Bedfordshire. Competition between the two companies developed into a real race against the clock. Would the private company be more apt to deliver a reliable and efficient product in a shorter period? The stakes were high. The best performing dirigible could well obtain the contract for flights between Great Britain and her Empire in India. The team of the *R-100*, in 1929, was less than optimistic concerning its chances of being the first to make the trip to India, with only one refuelling stop. Morpugo, in his biography of Barnes Wallis, designer of the *R-100*, cites Wallis concerning the *R-101*: "It is the crudest piece of design which I have ever seen. It is in fact inferior to the framework of Naval Ship nº 1…"

The contest was somewhat unequal. The government's company was allowed a big budget; and, moreover, would a minister dare to disown a company arising from his own jurisdiction? At least, the exercise seemed to give the appearance of a degree of equity between the private and public enterprises. Qualified as being the "capitalist" airship, the plans of the *R-100* had been entrusted to Neville Barnes Wallis, an employee of the Vickers Company. Born in 1887, in Ripley, in Derbyshire, and son of a doctor, he was recruited by Vickers to design airships. Along with H.B. Pratt, he conceived the *R-9*, *R-23* and *R-26*. He was even the Chief Engineer of the *R-80*. Later, he would create the Wellington bombers. The first work, very theoretical, started in Vickers'offices, in Westminster, in the year 1924. The plans were conceptualized a few months later at Crayford, in the Kent region. The evolution of the plans required many calculations. Since computers did not exist at the time, reliance was placed on the talent of a team of "calculators". "Calculator-in-chief" was the novelist Nevil Shute Norway (1899-1960). He was the author of Lonely Road (1932), Ruined City (1938), A Town Like Alice (1950), and On the Beach (1957). He was part of the crew that made the trip aboard the *R-100* to Saint-Hubert (in Canada). Norway was assigned the task of calculating the resistance of the airship relative to the total weight of the airship in motion. A very delicate operation that could lead to success or failure.

The members of the design team for the *R-100* moved to Howden on April 6, 1926. The workshops were in fact five kilometres away from the village of Howden. About twenty small houses were built near the hangar, to lodge a few of the employees. In 1925, the company had inherited a rather shabby site. An entire year had been required in order to transform it into a respectable space centre. Near the hangar, a factory capable of producing hydrogen also had to be built. Constructed in 1916, the installations were abandoned in 1921. Also, the hangars were somewhat narrow for the construction of this aerial Leviathan. Nonetheless, they were at that time the biggest workshops in the world, 228.6 metres in depth by 45.7 metres in width. A few strikes over claims for better salaries caused some delays. Once the installations were completed, in 1927, production kicked off. A large part of the workforce was local. Howden started experiencing unexpected economic growth.

Other workers were hired, including women, and many settled in the village. The workweek came close to 50 hours. The weekly salary for foremen was around £4, double that of employees.

From July 30 to August 9, 1929, small gasbags were filled up with hydrogen. Suspended from a beam like curtains, by injecting hydrogen they took form. Hydrogen from a nearby hangar passed through underground pipes to arrive directly at the airship. The gas passed through the upper part of the small balloon's structure. Two of these were being filled each day. This was a fairly delicate operation because, unlike a gasoline pump, the flow of hydrogen could not be controlled; the flow itself could not be slowed, much less stopped, without risking an explosion in the factory. When all the small gasbags were filled, the airship floated by itself above the ground. The fifteenth gasbag however, got filled only on the eve of the first test flight.

The testing of the engines during the month of September terrified some employees. The noise inside the hangar was infernal and deafening. We must not forget that barely one meter separated the roof from the upper part of the airship. Communications therefore, had to be in writing or by gestures.

The construction of the *R-101*, the "socialist" airship, proceeded in parallel, without any sharing of information between the two teams. With more substantial financial means, the *R-101* team, wanting to put on more of a show, added more equipment despite the weight and limited carrying capacity of the airship. The British media coverage favored the *R-101*. But the *R-100* team was also resourceful. So it was that during a visit by MPs of the British Parliament to the production factory in Howden, a female employee was confided the task of serving tea to the dignitaries gathered in one of the rooms used for assembling the dirigible

The construction of the *R-100*, like that of the *R-101*, took a long time. Expected for 1927, the dirigible was only delivered in December 1929, five years after its initial conception. In Cardington, a lot of research contributed to delaying the release of the *R-101*, while in Howden, the lack of skilled personnel and strikes considerably slowed development.

SAME CONTRACTS, DIFFERENT RESULTS

The specifications for the construction of the two airships were the same during the submission phase of the contracts: maximum speed of 110 km/h, capable of carrying 100 passengers, maximum weight 90 tons, capacity of 141,600 cubic metres of hydrogen, capacity for lifting 160 tons including 60 tons of material. However, as a result of many modifications along the way, two relatively different airships were put into operation. The name *R-101* had not been randomly given to Cardington's craft. It was meant to show that it would be a more modern version of the *R-100*, considered as the latest model of the Zeppelin influence. Thus, the *R-101* would benefit from the latest technological innovations. For instance, it was equipped with very heavy diesel engines that had not yet proved their worth in the air. The *R-100* team had foreseen this possibility, but had dismissed it, estimating that the airship's overall performance would be affected.

The correspondence filed in the public archives of the Public Record Office of Kew, in England, informs us that a few British leaders believed it necessary to imitate other countries and to give a real name to these crafts. The Secretary of State for Air was assigned this responsibility. The King wished for names of constellations or such designations as Castor or Pollux for the *R-100* and the *R-101*. Finally, it was conceived that the *R-100* would be named Canada, but only upon its arrival in that country. This never came to pass.

Installation of the first elements of the structure of the *R-100*.
Louis Emard Collection.

Structure of the airship *R-100*. On the two photos above, note the two men on the rings during assembly at Howden. Photo below: two rings on the ground.

The central ring is surrounded with unfilled gasbags. The outer covering is in place.
Société historique et culturelle du Marigot Collection.

Two workers inside the structure of the *R-100*.

A team completes the installation of the pod housing two Rolls-Royce engines.
Times Wide World Collection.

A team completes the installation of the pod housing two Rolls-Royce engines.
Times Wide World Collection.

The airship suspended from many cables. Note the ladders on the left of the photo.
Times Wide World Collection

Vickers Collection.

A few workers, on a longitudinal beam of the *R-100*'s roof, add the finishing touches to the external part of the airship.

The entry to the rooms and the gallery is in the upper right corner; the dining room is located in the lower right corner.
Michel Pratt Collection.

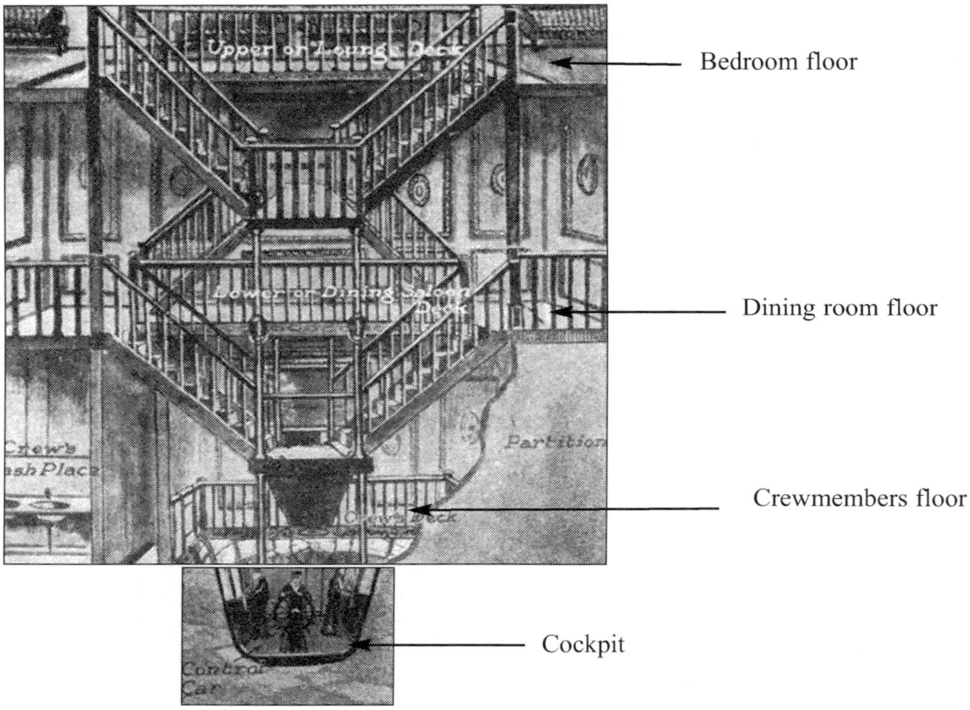

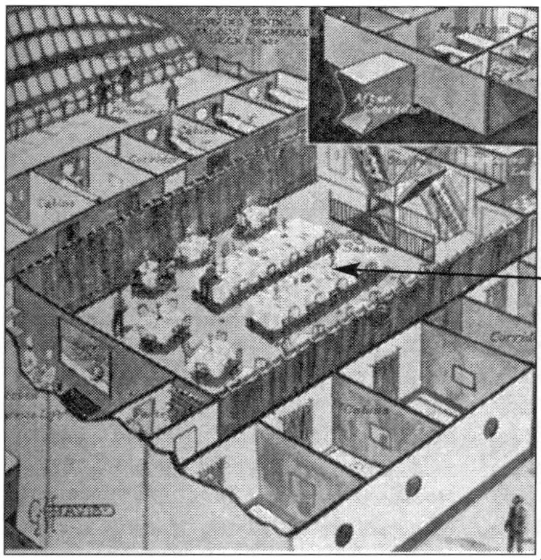

Living and dining rooms
Vickers Collection

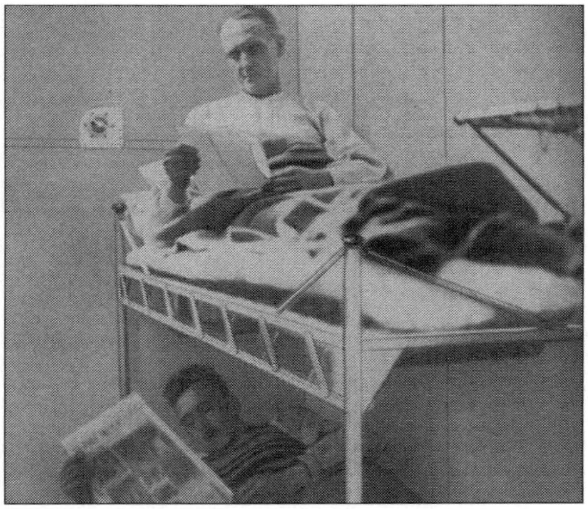

Bedroom

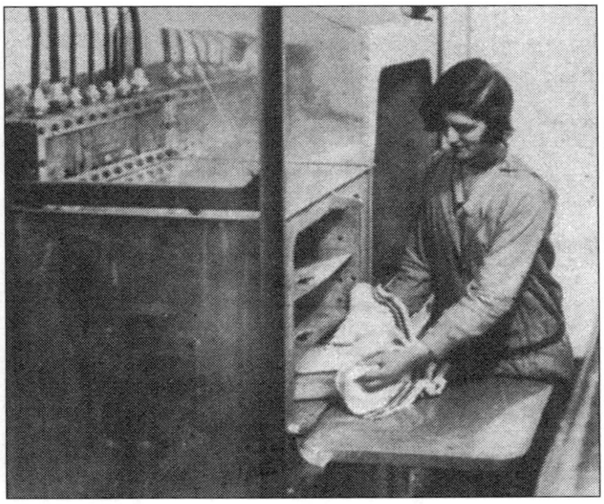

Kitchen

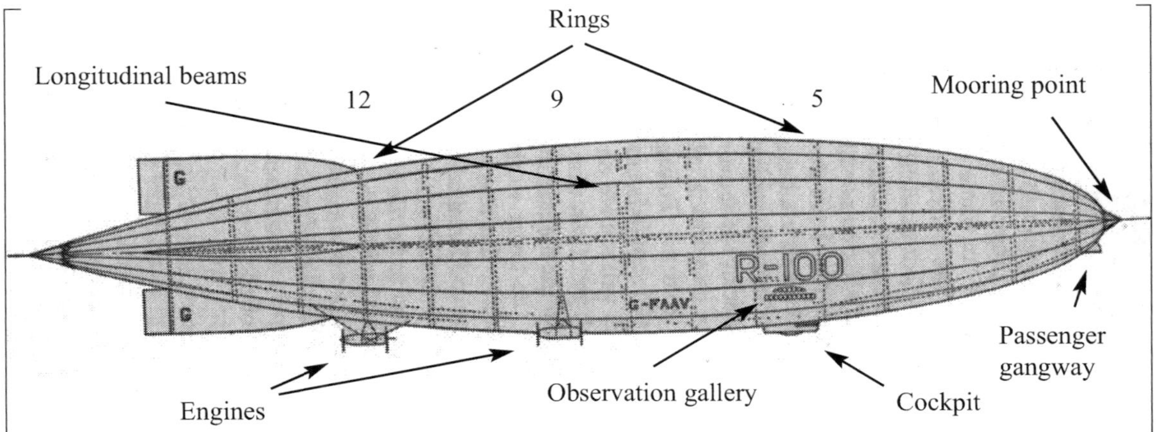

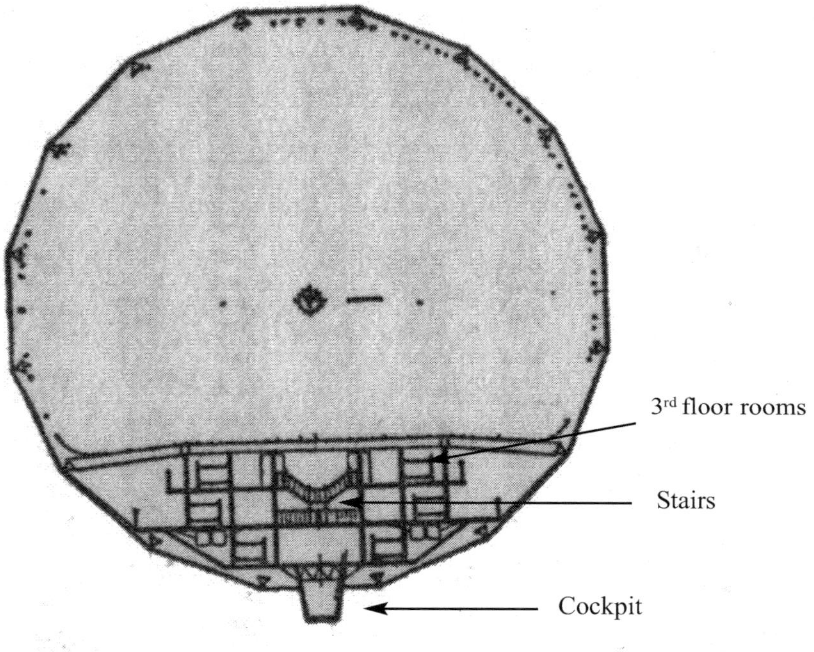

Ring nº 5 front view

CHARACTERISTICS OF THE R-100

From the tip of its blunt nose to its rather dynamic tail, the *R-100* measured 216.1 metres with a diameter of 39.6 metres. The frame was composed of 15 parallel rings; each ring in the shape of a polygon of 16 sides, was attached to 16 longitudinal duralumin girders made from an aluminum and cupper alloy but slightly bowed for the most part. These girders took on a triangular shape. One of these girders crossed the structure right through the center. The metallic structure was covered with a fabric stretching over 30 surfaces. Four ailerons were attached to rings 13, 14 and 15, at the rear of the dirigible.

While being assembled, the entire structure was suspended by cables to the hangar's ceiling. Especially during storms, the pressure placed on the old roof of the hangar often terrorized the employees.

Inside, the 15 gasbags or small balloons, made in Germany, contained 146,000 cubic metres of hydrogen and took the shape of the frame; they were protected by wire mesh and were suspended from the structure by rings. Each balloon was equipped with exhaust valves. These balloons were made in part from ox intestines impervious to the passage of gas called goldbeater`s skin. Imagine the number of oxen necessary to create the envelope for only one balloon! The storage capacity for hydrogen was the highest in the world at the time of *R-100* first flight; the *R-101* surpassed that mark.

The craft's structure was made from a light but resistant metal called duralumin. Duralumin itself is an alloy composed of 94% aluminum to which are added small quantities of copper, manganese and magnesium. The weight of a structure made of duralumin is equivalent to only the third of a structure made of steel. The structure was covered with a weatherproof linen fabric; the fabric was coated with a silver-colored aluminum paint. To attach the fabric to the structure, it had to be pulled from the inside by cables. The entire structure looked a bit like an umbrella.

The dirigible was propelled by six 12-cylinder Rolls-Royce Condor IIIB engines of 700 h.p. Quite a few times the company changed its mind about the choice of the engines. Initially they worked with engines that operated on hydrogen and kerosene, but realized later that the development process for these engines would not respect the fixed deadlines. They therefore settled for diesel engines; an experiment which lasted six months. However, senior management finally concluded that the engines were far too heavy, and they then turned to gasoline engines being used in aviation. Those engines were grouped in two in three gondolas suspended from the *R-100.* They were on the external part of the metallic structure to prevent all contact with the highly flammable hydrogen. The remanufactured Rolls-Royce engines were equipped with a reversible system: a considerable advantage over the *R-101* when the airship was moored or about to be moored.

The *R-100* depended on large tanks capable of storing 1,870 litres of gasoline each. Each tank could be activated by compressed air. In total, the *R-100* could store 15 tons of water: seven in containers made especially for that purpose and eight in tanks comparable to those for the fuel.

In order to gain altitude, the dirigible just had to jettison some of the water ballast, and to descend quickly, it only had to release some of the hydrogen. One section on the nose of the airship contained the cables necessary for mooring. The nose-cap, to fasten the cables to the mooring mast, was placed at the very end of the mooring cone. It was also through this same place that the passengers could get on board. On the roof, near the nose, an observation centre had been installed for the crew. The airship's tail supported its heavy ailerons as well as the rudder.

The craft was designed to carry 100 passengers with 22 kilograms of luggage each. The number of passengers could also be reduced to 50 in order to increase the weight available for freight. The *R-100* could cover in the air, non stop, a distance of more than 4,800 kilometres in less than 36 hours, if there was no wind. For instance, a trip to the British colony of India would have been broken into two phases of 36 hours each with a 12-hour stop-over in Egypt to refill the tanks.

The *R-100* constituted one of the first examples of the standardization of production. There were only 11 different pieces in the airship and only 15 different joints in the entire structure.

THE CREW'S QUARTERS

The crewmembers' decks were located on the first floor, along with the radio communications room and the kitchen, a room equipped with electricity. Meals were sent all the way to the dining room using a dumbwaiter.

The ship's control cabin was situated outside the dirigible's envelope, directly under the officers' chambers.

THE PASSENGER AREA

The dirigible could house close to a hundred passengers; they had at their disposal 14 two-berth cabins and 18 four-berth cabins, allocated between the second and third decks. All of these rooms were located primarily in the front of the craft, between rings 5 and 6, at a distance of 58 metres from the nose. The cabins themselves were only separated by a fabric wall. The soundproofing between the rooms was practically non-existent. In fact, they seemed more like tents than cabins. The bunks were placed in tandem. The dining room, situated on the second floor, could seat about sixty people. The walls, of a very light wood, were painted white. A double wall surrounded the entire passenger section. Air circulated between the walls to avoid the risk of flammable gas penetrating into the passenger section.

The passengers could admire the landscape from an observation gallery that stretched along a corridor on the first floor. In fact, the dirigible's envelope had been pierced to integrate a double row of stained-glass portholes. They were located directly under the letters *R-100* of the airship, just like the cockpit.

The two hangars of Cardington. n° 1, on the left, lodges the *R-101* and n° 2, on the right, lodges the *R-100.*

People could circulate through a long gangway, illuminated by electricity, which started from the nose, the embarking point, and stretched all the way to the passengers' quarters. On the third floor, a gallery surrounded the second-floor dining room with rooms on each side. Passengers were free to stroll around it.

THE OTHER SECTIONS

After the passengers' section the corridor narrowed considerably, having been designed strictly for the crew. About 40 metres behind the cabin section, one could find, on the exterior, two pods each containing two engines. Thirty metres farther, another pod of the same type enclosed engines. The engines were located far enough away so that their noise would not inconvenience the passengers. Then, a little farther on, there were the ailerons used for navigation and, finally, the airship's tail.

The two hangars of Cardington. no 1, on the left, lodges the *R-101* and n° 2, on the right, lodges the *R-100.*

VENTILATION

Ventilation was quite adequate and permitted the continual exchange of fresh air. If the air was too cold, the *R-100* would be heated by a radiator system using steam generated by two of the engines. Note that the airship's overall structure was composed of a double wall inside which air circulated to avoid the infiltration of toxic or flammable gases into the quarters lodging the crew or the passengers.

All these characteristics demonstrate that the airship was conceived to respond primarily to an international calling.

Mooring manoeuvres of the *R-100* at the Cardington tower. The arm at the head of the tower is slightly inclined and cables stabilize the craft. The pointed tail of the *R-100* indicates to us that it is the original version.
Flight Collection.

Test flight of the *R-100* above Farnbourough Airport, in Great Britain. A small airplane accompanies the airship.

Above England.

On the first photograph of the preceding page, we see cables that seem quite thin. We also have a terrestrial background for *R-100*. Here, the view from the air provides a new dimension to the scene, even if it is of the *R-101* and not the *R-100*.
Airship Heritage Trust Collection.

On April 26, 1930, the *R-100*, moored in Cardington, receives a courtesy visit from the *Graf Zeppelin*, as seen in the background.
Airship Heritage Trust Collection.

The *R-100* moored. Obviously, visitors preferred the *Graf Zeppelin* that could land directly on the ground, without using an mooring mast.
Société historique et culturelle du Marigot Collection.

On January 25, 1930.

The original version of the *R-100* in Cardington with its pointed tail is shown in the uppermost photo and the final version, with its new rounded tail section, in the other two photos.

THE FIRST TESTS

On the morning of December 16, 1929, the weather was ideal, there being no winds as a prerequisite, to take the *R-100* out of its very narrow hangar. A centre line had been painted on the ground to guide the 500 soldiers who were to move the airship. Sir Dennistoun Burney, Barnes Wallis, Reginald Colmore, Ralph Sleigh Booth, George F. Meager, Nevil Shute Norway and 51 others were inside the ship during its exit, tail first. Norway, who had not seen a thing, understood that the dirigible had been successfully removed from its hangar when he heard the cheers erupting from the crowd. The operation was under the direction of Major Scott, responsible for both the *R-100* and the *R-101*. He made sure that the airship was far enough away from the hangar before climbing aboard. Ralph Sleigh Booth, nicknamed "Mouldy", inaugurated the first test flight of the *R-100* just as he had done, back in 1916, for the first flight of a British dirigible in Howden. He jettisoned half a ton of water from the nose and another ton from its tail and then commanded that the ship be allowed to fly. At a height of 150 metres, two of the six engines were activated and thus started a slow flight over Howden. The captain steered the airship by using a big wheel of 1.2 metres in diameter, similar to the helm of a large naval vessel. As a safety precaution, there were parachutes for everyone on board.

While headed for Cardington, 210 kilometres from Howden, breakfast was served on board. In his biography, Nevil Shute Norway admitted he had been terrified by the possibility of a crash. He did not fear anything when the ship was in the air, but he worried about a sudden fall to the ground due to the huge risk of explosion. Norway confided to having brought with him a blade razor and a very sharp knife that he would have used to cut the fabric if the craft had ever caught on fire. When the *R-100* arrived in Cardington, after five hours and 47 minutes of flight, it was to use the mooring mast for the first time. There was only one for the two airships. If oneof them was moored, the other had to be stored in the hangar. The next day, the *R-100* made another outing, lasting six hours and 29 minutes, then, in the evening, it was housed in Cardington's hangar. Some repairs and adjustments were made to the dirigible; in particular they changed the 3,600 ropes that held the exterior envelope to the frame. As of the second week of January 1930, the *R-100* was ready to undergo some new tests. Meanwhile, back in Howden, the dismissal of almost the entire staff caused a major setback. The economic crisis, which had started in late October 1929, in the United States of America, began to make its effects felt in Great Britain in early 1930. Morale was quite low in the field of dirigible construction.

Nevertheless, further tests were carried out: first on January 16, a 13 hour and 20 minute flight took place, and on January 20, another one of seven hours and 10 minutes. Speed tests were conducted during the flight on January 16th, after which many small holes were discovered in the fabric covering the *R-100*, and particularly directly under the fuel tanks. Departing from Cardington at 9:10, on the morning of January 20, the *R-100* passed over Farnborough at 11:00. An airplane followed the ship and filmed the fabric's performance from all angles. This film would allow the team to pinpoint the weaknesses of the covering. The contract of the *R-100* called for an endurance test of 48 hours as well as a test flight within the British Empire, in India. Nevertheless, as the *R-100* team had already given up using diesel engines, preferring to use gasoline engines, it was

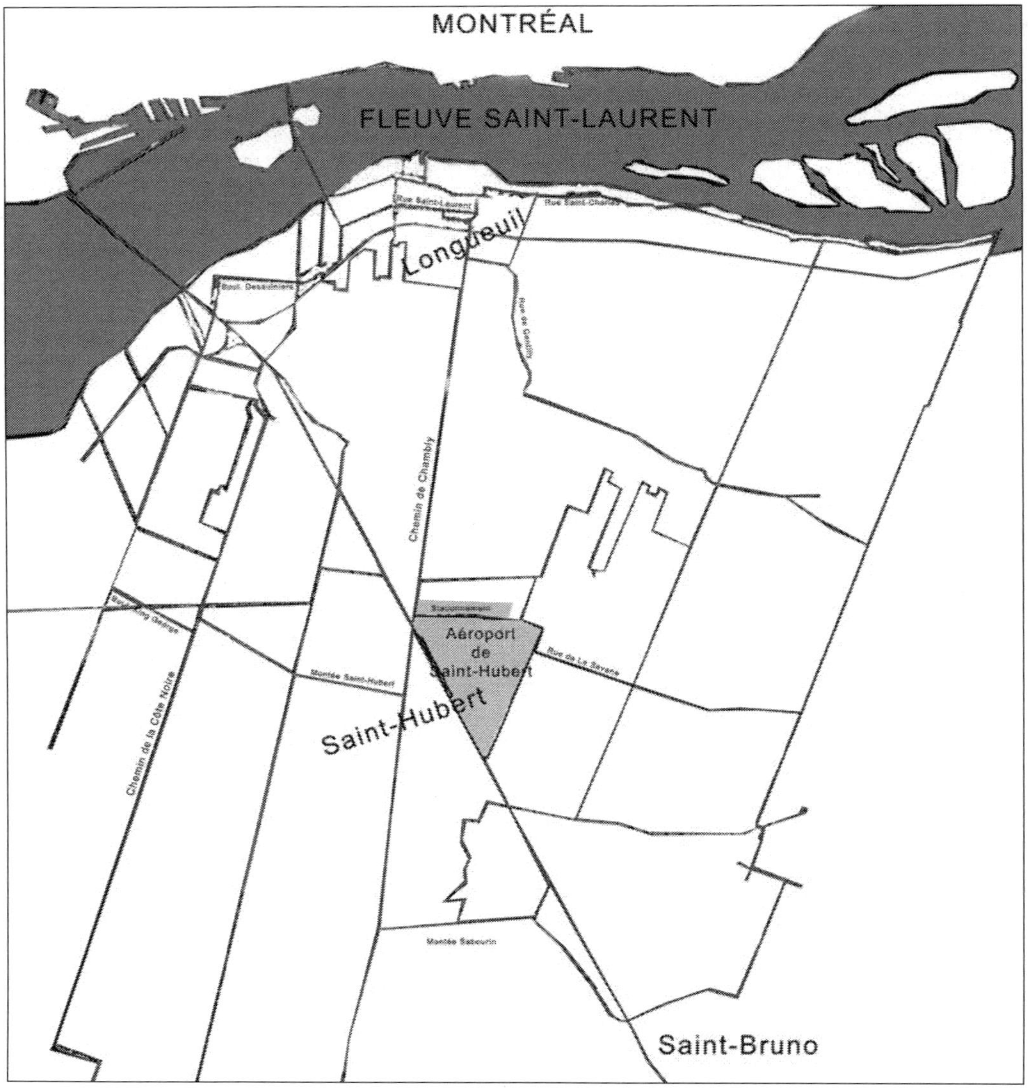

Site of the Saint-Hubert Airport.

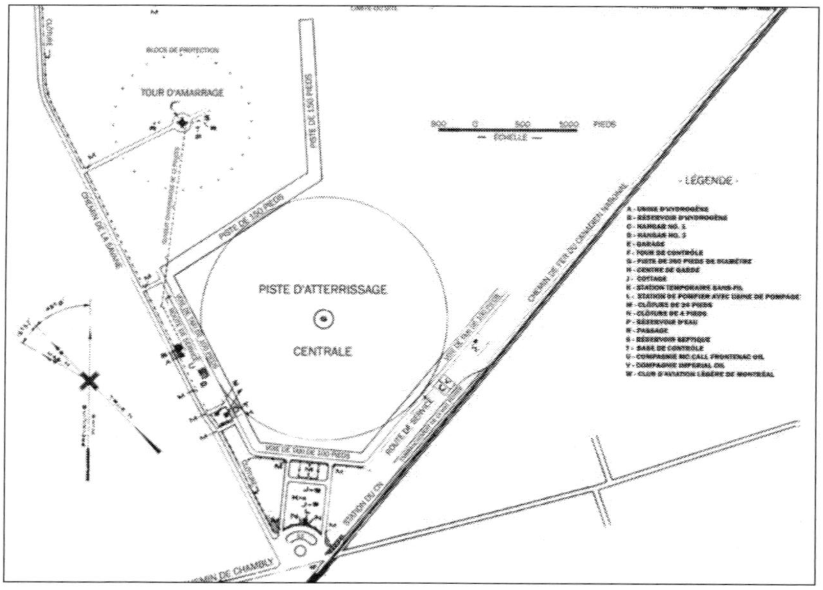

considered hazardous to fly over tropical regions with such engines. Hence, Canada was selected asthe place for the test flight. The test that was to last for 48 hours took 54 hours and 52 minutes instead, from January 26 to the 28th. For this fifth flight, it was planned to fly close to Spain and the Azores, but the Air Ministry gave formal orders to fly only over the United Kingdom. At the end of January 1930, 87 hours of flight time had been accumulated and the ship had stayed at its tower for more than 305 hours. At the end of the month of April, the *R-100* was moored to the Cardington tower.

On April 27, the *LZ-127 Graf Zeppelin* came to Cardington for a courtesy visit. The crew members and the technicians from both sides fraternized for a few hours while a group of people seized the opportunity to admire, at close range, the formidable German airshipwhich had made a world tour in 21 days and which, measuring 265 metres, was then thelongest in the world.

On May 22, 1930, the *R-100* took the residents of Manchester by surprise. The airship was flying so low that the roar of the engines made many people come out of their houses. Some even climbed onto the roofs of factories and offices to witness the spectacle. They could clearly see the letters and the numbers written upon the dirigible. The sun's rays turned leaden the silvery airship. The people, gathered for the most part near the town hall, greeted the crewmembers. The dirigible then headed towards Salford. Gradually, the roar from the engines subsided. The clamor from the excited throng then quickly took over. Despite its disappearance in the sky, people stood and waited for long periods of time, as if expecting the flying marvel to reappear. That sixth test flight, of about 20 hours, had been planned to test the reliability of the Rolls-Royce Condor IIIB engines and to measure fuel consumption. The ship had damaged its pointed tail, bent at a 45° angle, forcing it to prematurely return to its hangar after about 20 hours of flight. The tail was fixed by rounding it and by reducing its length by close to five metres.

On July 25, the *R-100* made a seventh and final 24-hour flight before it's crossing of the Atlantic. The results were interesting. The *R-100* reached much higher speeds than those stipulated in the contract; on the other hand, it failed to lift the weight prescribed in the same contract. As for its construction, the firmness of its exterior coating still provided headaches. Regardless, it was decided not to house or store the *R-100* in the hangar but to leave it moored to the tower until its departure for Canada.

THE CARDINGTON HANGARS

Before 1929, the *R-100* was housed in Howden, in a hangar opened on June 26, 1916. Howden also possessed two smaller hangars for non-rigid airships. This hangar which served for the *C-4, C-11, C-19* and *C-21* measured 228.6 metres in depth by 39.6 metres in width.

In 1930, the two airships, the *R-100* and the *R-101*, were accommodated side by side in two Cardington hangars, 80 kilometres away from London, where the *R-101*was built. The *R-101* hangar, erected in 1917 by A.J. Main & Co., had also served for the *R-31* and the *R-32*. It was enlarged between 1924 and 1926 by Cleveland Bridge & Engineering when its depth passed from 213 to 248 metres, and its height was increased by 11 metres.

Mooring tower at the beginning of its construction, during the summer of 1928.
Louis Emard Collection.

Saint-Hubert mooring tower. On the top, the beginning of the telescopic steel arm designed by the British firm Babcock & Wilcox can be seen.
Louis Emard Collection.

The extendable arm which was to receive the nose cup of the *R-100*.

Mechanism of the telescopic arm seen from a different angle.

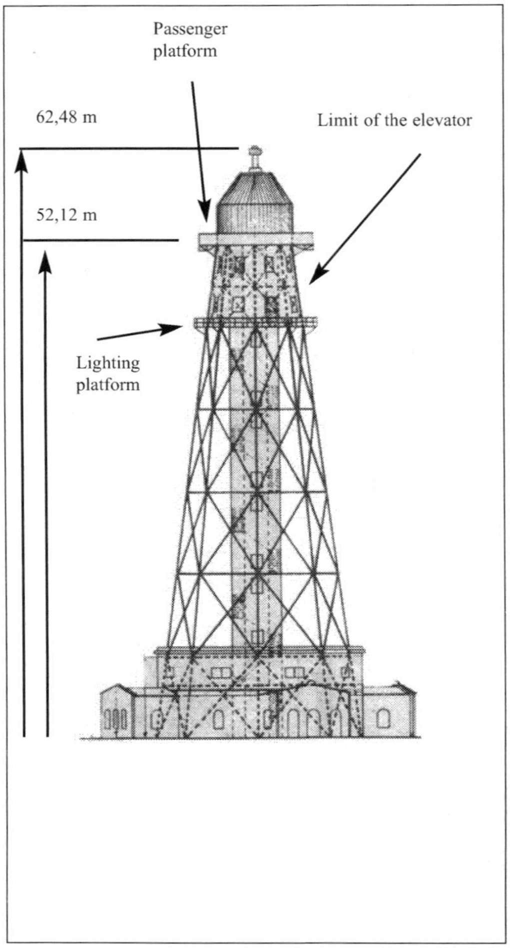

The mooring tower under construction.
Michel Pratt Collection.

The Saint-Hubert tower under construction with the peripheral installations.
Michel Pratt Collection.

The second hangar, for the *R-100*, came from the dismantling of the hangar from the airship base at Pulham Market, Norfolk; it had cost £150,000 in 1919. The wooden hangar was also reconstructed and enlarged. All have been preserved for future use. Hangar number 2 that housed the *R-100* was also renovated in 1993 by the Building Research Establishment, owner of the site. Four tons of paint were used in that renovation. Each hangar had an area of five acres or 736,200 cubic metres of space. The dimensions were: 248 metres long, 84 metres wide and 55 metres high. The doors were made of steel and took more than 15 minutes to close. One hangar would have accommodated 16 Olympic swimming pools.

The Cardington hangars in the 1990s. Photo by Marc-André Valiquette.
Louis Emard Collection.

:

Radio installations at the Saint-Hubert air base. Note that the employees of Bell, in the centre of the photo, have highlighted their company's logo.
Library and Archives Canada Collection, C 64910.

The CN had built this wooden tower of more than 18 metres beside its railway line to broadcast throughout its network the arrival of the *R-100*.
Library and Archives Canada Collection, PA 122172.

PREPARATIONS FOR THE VISIT OF THE R-100 IN CANADA

The Canadian Government had committed itself, at the Imperial Conference of 1926, to contribute to the desire of the British Empire to serve as the cornerstone of air transportation, by constructing a mooring mast and developing an adequate meteorological information or weather forecasting system. Prime Minister William Lyon Mackenzie King could not have been clearer: "May I, first, say in regard to the specific request which the Secretary of State for Air has made that the Canadian Government will only too readily cooperate with the British Government in the way he has suggested, by immediately taking steps to see that mooring masts to secure the landing places for airships in Canada are erected: also that the work of meteorological organization is commenced forthwith." He went as far as to point out that Canada could provide the helium: "Possibly Canada will also be able to assist by provision of helium gas…"

Helium was a commodity in great demand because it is a gas that is lighter than air (1/7). Moreover, it is non-flammable and can even contribute to extinguishing a fire. The United States, however, had the monopoly on this source of energy found in the states of Texas, Kansas, Wyoming, Utah and Colorado. There was even a law prohibiting its export. As soon as the Estimates for 1927-1928 were voted on, a budget allocation was included for the purchase of an airbase.

THE ANCHORING MAST

In Canada the first preparatory step for the coming of the dirigibles was to choose a location for the mooring tower, which, in its form, resembled an enormous lighthouse. Its function was the same as that of a dock for a boat. Once moored to the tower, the dirigible floated lightly in the wind just as a boat is gently swayed by the waves. Normally, whe not flying, a dirigible would be moored to a tower. It is only when a major overhaul was needed that it would be returned to its hangar.

In the spring of 1927, the mooring masts of Cardington, in Great Britain, and of Ismailia, in Egypt, were almost completed, while the one for Canada had not even been started. The mooring mast of Cardington had been built at a cost of £127 000 by the Cleveland Bridge & Engineering Company. The upper part of the tower had, however, been designed and built by Babcock & Wilcox.

On April 22, 1927, Herbert Scott, Flight Supervisor for the *R-100* and A.R. Gibbs, of the civil engineering department of the Works Division of the British Air Ministry, arrived in Montreal. On May 3, they met with the Minister of National Defence of Canada, J. L. Ralston, who tried to convince them to choose between one of two sites near Ottawa, the Nation's capital: the Connaught Rifle Ranges, a military training center located to the west of Ottawa, facing the Ottawa River, and Rockcliffe, it too fronting on the Ottawa River and even closer to Parliament Hill. Visits to these locations, on May 4, were inconclusive. The two British spokesmen had already visited Saint-Hubert upon their arrival, and their choice was made. Nevertheless, they visited some sites in Nova Scotia and in New Brunswick as well as in Valcartier, near Quebec City. But, at the end of the month of May, they finally settled upon Saint-Hubert and its flat and triangular-shaped land of 210 hectares, near the Saint-Hubert train station. The triangle was bordered by Chambly Road, de la Savane Street and the tracks of the Canadian National Railway. On June 24, 1927, the acquisition of the land was authorized by Order-in-Council 1233, at a cost of $177,000, duly purchased in the month of August.

The mooring mast system was designed to facilitate the mooring and to avoid having to wait for ideal weather conditions in order to put the craft into service. Nonetheless, Charles Dennistoun Burney himself found the system technically inadequate and economically inappropriate.

During the month of July, the company Babcock & Wilcox was given the job of building the mooring tower's cap. The British had strongly insisted upon the need to have mooring towers identical to those used in the Cardington tests. The top-most portion of the mooring tower was delivered in August 1928. The construction of the octagonal steel tower, undertaken by Canadian Vickers, was completed during the autumn of 1928. The tower was surrounded by a platform for lighting fixtures located at a height of 43.5 metres and another platform for passengers at 52 metres, of a diameter of seven metres, while the base of the tower measured more than 21 metres in diameter. The lighting had been installed with the collaboration of the Canadian General Electric Company. Each of the eight poles of the tower was fixed on a concrete base measuring 3.3 metres by 3.9 metres by a depth of 2.9 metres.

The detailed plans for the *R-100*'s landing dock not having been delivered by the British Air Ministry, the builders realized, only a few days before the *R-100*'s departure, that they needed to build a wood platform to compensate the 60 cm gap left between the dock and the footbridge of the mooring tower. The cruciform building under the tower was different from Cardington's. It was composed of different modules distributed on two floors. Furthermore, it was incorporated inside the tower in a closed structure. In Cardington, where the elevator was built into an open-air structure, one could see the elevator moving, but not in Saint-Hubert. At the top of the tower there was a telescopic arm, extendable and retractable over two metres, which could be adjusted at the nose of the airship. The arm weighed more than 15 tons and could resist a pressure of more than 40,000 kilograms. It could be inclined to an angle of 30 degrees. Operated by compressed air when the ship's nose was attached to it, the arm lifted up vertically, retracted and locked itself, still allowing the dirigible to move freely around the mooring mast and to be able to make, at least theoretically, a 360° turn. All of the important cabling was controlled

Charles Dennistoun Burney, the "builder" of the *R-100* and Major George Herbert Scott.
Flight Collection.

Charles Dennistoun Burney, the "builder" of the *R-100* and Major George Herbert Scott.
Flight Collection.

by a device that relied upon the electro-hydraulic principle. The three mooring cranks were activated by variable speed gears.

The tower stood 62.5 metres, excluding the arm's extension; from the air it looked like a spider web. It was glossy orange and black in colour, whereas the building, at the base of the tower, was coated in grey stucco. On the passenger platform, a mobile staircase adjusted to the dirigible's position. Passengers could disembark from the dirigible using a small and slightly inclined gangway, to then take the elevator capable of holding up to 1,360 kg and which travelled at a speed of 45.7 metres per minute. Thomas Dunlop Rankin had been the architect assigned to theconstruction of the mast.

THE FUEL

A tank containing 10,000 gallons of fuel, buried underground, easily sufficed to supply the airship's Rolls-Royce engines. This tank was connected to the water distribution service. It was the water pressure that pumped the fuel back up to the surface through a pipe of approximately five centimetres. A system of valves assured security.

THE WATER

The water supply was assured through the collaboration of the Canadian National that agreed to the installation of two long pipes along the rail line to its twin reservoirs at Southwark, situated about 3.2 kilometres north of the tower. A 20 centimetre pipe ensured the delivery of water for the dirigible, and another one of 10 centimetres, for domestic consumption. At the Saint-Hubert airfield, the water was stored in a cement reservoir holding up to 50,000 gallons. The pumps and the filters were in a brick building, near the reservoir.

THE HYDROGEN

Hydrogen was essential to fill up the balloons. A 1,428 cubic metre gas reservoir was also necessary. Two hydrogen generators, each able to produce 571 cubic metres per hour, were installed in a building near the reservoir, at a distance of 762 metres from the tower.

THE TELEPHONES

The Bell Telephone Company had installed numerous telephones almost everywhere: in the mooring tower, the administration offices, the hangars, the weather centre, the first-aid centre, the control tower, and in the buildings that housed the machinery and the reservoirs. Many other telephones and a telegraph service were also installed in the press-room. Even on the field, Bell had installed public phones for general use.

THE METEOROLOGICAL DATA

The coming of the *R-100* caused turmoil to the transmission of meteorological data in Canada. It was the Canadians' responsibility to ensure communication links with the *R-100* from the instant it crossed the 35° of longitude. Consequently, the new Saint-Hubert weather station stayed open 24 hours a day during the trip, also guaranteeing a telegraph service. The meteorological and flight data were transmitted by different Marconi stations. The Marconi station of Bodmin, in Great Britain, sent the first reports which were received at the Marconi station of Yamachiche, in Quebec. In Eastern Canada, it was the Marconi station of Louisbourg, in Nova Scotia, that ensured intermediate transmission. In Quebec, it was the Canadian Marconi station in Drummondville that transmitted return messages to the one at Bridgewater, in Great Britain.

THE RADIO LINK

It was crucial to be able to communicate well between the *R-100* crew and a team on the ground. During the arrival of the *R-100*, a few technicians from the Transmission Service climbed on board to install a radio-wave beacon receiver. It was said that the British crew doubted the precision of the Canadian receiver and even bet a bottle of Scottish "spring water" that the airship could not be guided over the Peace Tower, in Ottawa, by that radio receiver. Yet the trip went off perfectly and the British crew kept its promise. The quality of the receiver was so good that it was even kept for the return trip to Great Britain.

LOCAL TRAFFIC... PRIORITY TO TRAIN TRAVEL

The Canadian National Railway (CNR) set aside 10 trains per hour for all the visitors. Each train had 15 cars able to carry 1,200 passengers. The CNR even built some additional tracks west of Guy Street, in Montreal, to avoid parking problems. More than 250 individuals worked day and night to meet the deadlines. Two platforms of more than 300 metres were also built.

The opening of the du Havre (later to be named Jacques Cartier) Bridge really sped up the crossing from Montreal to Longueuil. Victoria Bridge tickets could even be exchanged for those of the du Havre Bridge.

THE COMMUNICATIONS SERVICE... A SCANDAL! ACCORDING TO FRENCH-LANGUAGE GROUPS

The communications network for the *R-100* created a wave of protest from Frenchspeaking groups in the country. The daily La Presse threw the first stone, claiming to be outraged that the live broadcasting of the event had been assigned to two unilingual English-speaking journalists. The Controller of Civil Aviation of Canada, J.A. Wilson, had indeed declared his intention to delegate the event's coverage from the mooring tower to T. Shields and Thomas D'Arcy Finn, both of the Ottawa Evening Citizen. In short order, the newspaper received support from the Société Saint-Jean Baptiste, students from the Université de Montreal, the Canadian Alliance for the Vote of Quebec Women and the

Action Catholique. The Minister of National Defence, Colonel John Layton Ralston, allowed a French-speaking journalist, Jacques-Narcisse Cartier, from the CKAC radio station and La Presse, to be present as well. Mr. Cartier was just recovering from two years spent at his house in Saint-Gabriel-de-Brandon suffering from tuberculosis.

THE OFFICERS AND CREW OF THE R-100

The six officers of the *R-100*

Ralph Sleigh Booth (1895- ?). Born in Hughenden, Bucks, Captain of the *R-100*; he had carried out the first test flight for the *R-100*. Experience in the *R-24* and the *R-33*.

Ernest L. Johnston (1891-1930). Born in Sunderland. Squadron Leader, excellent navigator. He had been employed in 1924 by the Royal Airship Works. He had also worked for Imperial Airways. Navigator for the *R-100* as well for the *R-101*. Killed in the *R-101* accident.

George F. Meager (1893-1976). Born in Newport, Isle of Wight. First Officer. Renowned for being very meticulous. Author of My Airship Flights. He was a Captain in the Royal Naval Air Service.

George Herbert Scott (1888-1930). Born in Catford, Kent, responsible for flight supervision. He had been a captain for the *R-9*, the first rigid British dirigible to fly. Captain for the R-34 during the first crossing over the Atlantic in July 1919. Major in the Royal Air Force, he had designed the British dirigible anchoring system. He made many tests on the *R-100* and the *R-101*. He was the one who fixed the exact time for the takeoff, the course to follow, the speed and the altitude of the dirigible. Major Scott had become a very controversial character, with a reputation for making unfortunate decisions. Captain Ralph Sleigh Booth also put his foot down. Scott was not to give any orders to crewmembers and was to act only as an advisor to the captain of the *R-100*, a warning he was not to respect.

Maurice H. Steff (1896-1930). Born in Luton, Beds. Second Officer. Employed as Second Officer for the *R-101* in 1929. Killed in the *R-101* accident.

Maurice A. Giblett (1894-1930). Born in Englefield Green, Surrey. Meteorologist. Experience in the *R-80, R-36* and *R-38*. Killed in the *R-101* accident.

The crew

Name	Age	Assignment
Angus, W.Y.	41	Chief Engineer
Atkins, G.K.	30	Wireless Radio Operator
Ball, R.	31	Engineer-Chargehand
Broughton, C.	28	Rigger
Clarke, H.W.	34	Engineer-Chargehand
Cumley, H.	24	Engineer-Chargehand
Cutts, C.G.	27	Rigger
Deverell, R.L.	27	Rigger
Disley, A.	28	Radio operator
Eldridge, A.	35	Captain's Clerk
Flatters, C.	25	Rigger
Gaye, R.	24	Engineer-Chargehand
Greenstreet, T.	33	Chief Coxswain Flight Sergeant
Hobbs, T.	32	Assistant Coxswain
Hodnett, F. (Curran)	28	Assistant Steward
Hunt, L.A.	24	Pilot
Jowett, J.	34	Engineer-Chargehand
Keeley, S.T.	35	Chief W/T Operator
Lelliot, D.	29	Engineer-Chargehand
Long, G.E.	36	Assistant Coxswain
Mann, N.	37	Engineer-Chargehand
Meegan, J.F.	30	Cook
Megginson, T.W.	18	Cabin Boy
Millward, H.	24	Engineer-Chargehand
Moncrieff, L.A.	32	Assistant pilot
Rumsby, C.H.	28	Rigger
Savidge, A.H.	32	Chief Steward
Scott, G.R.	30	Rigger
Stupple, E.J.	34	Engineer-Chargehand
Sturgeon, J.M.	38	Engineer-Chargehand
Watts, G.	31	Engineer-Chargehand
Williams, F.	43	Rigger
Wiseman, A.F.	27	Rigger

The jackets of the officers were adorned with gold buttons while the rest of the crew contented themselves with black ones. A symbol representing a crown, as well as the inscription "*R-100*" and "Royal Airship Works, Cardington" were sewn on the caps of the crewmembers. As pointed out by James Leasor, in his book, *The Millionth Chance*, the influence of the navy on the organization was very strong. After all, was not the dirigible an airship?

Among the passengers, there were: Reginald Blaney Bulteel Colmore, Director of Airship Development; Archie H. Wann, Squadron Leader, former captain of the *R-38*; Frederick M. McWade, Inspector of the Aeronautic Division of the Royal Airship Works;

Dennistoun Burney, Managing Director of the Airship Guarantee Company; Nevil Shute Norway, Chief Calculator of the Airship Guarantee Company; Lieutenant Commander R. St. J. Prentice, representative of the British Admiralty.

The biggest surprise was certainly the absence, on that historical transatlantic crossing, of the ship's designer, Barnes Wallis. From January 1930, the Vickers Company transferred him to another area and insisted, in spite of Wallis' request, on keeping his "genius" in Great Britain. Obviously, Wallis and Burney no longer got along at all. Since 1926, the year Wallis came to settle in Howden, relations had not ceased to deteriorate between the two men. Wallis claimed that it was HIS airship while Burney rather insisted it to be HIS. It should be noted that the official reports do not reveal the presence of a young assistant-steward, T.W. Megginson, who had barely turned eighteen. The same mistake was made, according to George Meager and Geoffrey Chamberlain, concerning another assistant-steward, F. Curran. From all appearances, Curran may have also used the name Hodnett.

The *R-100* passing above the Mersey River, near Liverpool, and heading towards Canada.
Illustrite Zeitung, nº 4456.

Above Guernsey.

THE R-100' S TRIP TO CANADA

Before the *R-100*'s departure, its tanks had been filled with 47,279 litres of fuel plus 5,000 kg of ballast. There were also 400 kg of food and 1,800 litres of water. In almost total anonymity, the airship left the Cardington mooring tower on July 28, 1930 at 9:48 p.m., Montreal time. Part of that anonymity can be explained by the fact that it was 02:48, Greenwich Mean Time, at departure. The ship therefore, had left during total darkness.

At 40 metres in diameter, the *R-100* was the world's largest dirigible when compared to the 30 metres of the *Graf Zeppelin*. It was shorter than the Graf Zeppelin, which measured 54 metres, but its gas storage capacity was greater: 141,600 cubic metres, versus 104,800 for the Zeppelin.

It was George Meager who was at the controls. Captain Ralph Sleigh Booth was supervising. During its departure, the *R-100* had released 1,800 kg of ballast. At 23:00, Meager went to bed dressed in “Teddy Bear” pyjamas, only to be woken up two hours later in order to replace Booth as flight supervisor. The ship had already flown over the towns of Wellingborough, Northampton and Rugby.

The airship flew over the Roman-Saxon town of Chester on Tuesday July 29, at 1 o'clock in the morning, and then the crew changed course flying directly over Liverpool where train and steam-boat whistles were heard. The pilot was very impressed by the size of the Anglican cathedral, the biggest in Great Britain, with its 100 metre high tower. Still under construction at the time, the cathedral was not completed until 1978. The captain made the decision to fly over Ireland in order to avoid the unfavorable atmospheric conditions that were raging at that time over the ocean. The trip across the Atlantic was made without difficulty. On July 30, at 02:10, Montreal time, the airship met the SS Ausonia. There was only a small inconvenience: the temperature was rather cold on board, especially in the Newfoundland region. The ship was easier to get into to the air than to heat, considering the rudimentary systems available.

During nine hours, over a distance of 960 kilometres, the *R-100* could not make any sort of direct communication with either Canada or Great Britain. The dirigible could nevertheless communicate with ships such as the Winchester Exporter, the Arabic, the Caledonia and the Montclair. During many long hours, the earth's surface could not be seen from the *R-100*; in fact, not since it had flown over Oversay Island, at the north end of Great Britain, near the northern part of Ireland. The 46 passengers were almost delirious with enthusiasm when they could finally make out the Cunard liner, Ausonia. Numerous communications were exchanged between the two crews. The dirigible passed over the Strait of Belle Isle, through a thick fog; visibility was nil. The *R-100* confronted one storm over Anticosti Island at 8:30, and another one, over Pointe-au-Père, at 11 o'clock. Meanwhile, the crew had noticed the presence, on the St. Lawrence River, of the Empress of Scotland as well the Duchess of Bedford. Around noon, they noticed that a 4.5 metre part of the covering on a tail section had been torn off by the wind. They spent two hours repairing it. From that point, it became less and less probable that they could meet the goal of reaching Saint-Hubert early in the evening. The airship flew slowly and at a very low altitude over the city of Quebec on Thursday, July 31, at 6:00 p.m. People had gathered in parks and on footbridges to admire and acclaim the airship. Later in the evening, around 7:40 p.m., the crew encountered a violent storm in the region of Three Rivers. The ship's nose pitched toward the ground at an angle of more than 30 degrees. Tables in the dining room got knocked over. The Chief Steward, Mr. Savidge, was copiously soaked by spilled soup, and then the ship gained height and rapidly reached an altitude of 1,200 metres. Major Scott, commander-in-chief on board, challenged the storm rather than going around it. The left aileron was damaged, on the side of the *R-100* lettering; the fabric was torn by the wind at two places over a surface of about four metres by two metres. In his biography, Nevil Shute Norway questioned Major Scott'sstrategy. He mentioned that Scott later also ignored the weather with the *R-101*, leading it to a catastrophe.

Around 1:10 a.m., the crew rejoiced at the sight of Montreal's illuminated skyline, yet it took them some time to understand the meaning of the cross in lights placed on Mont Royal. In Saint-Hubert where many curious bystanders were already waiting, the uproar was indescribable: sirens and car horns each tried to drown the other out. Spotlights projected beams of light on the *R-100*. Officers of the RCMP (Royal Canadian Mounted Police), on horseback, tried to clear the way.

The *R-100* made a long detour, all the way to Pointe-aux-Trembles and then towards Westmount, before finally returning to Saint-Hubert. The crewmembers delayed its entry to Saint-Hubert because they were not familiar with the airport, especially the location of the high voltage wires. The officers waited for the sky to brighten up a bit before proceeding with the mooring.

On Friday August 1st, at 4 a.m., the dirigible released the guide rope and attached itself to the orange and black mooring mast. The tower, 62 metres in height, was illuminated. The trip had lasted 78 hours and 49 minutes; the airship had flown therefore, a distance of 5,414 kilometres at an average speed of 67 km/h. It should not be overlooked that only a few planes had managed, at that time, the feat of crossing the Atlantic Ocean from Europe to America, without any stopover. Despite the delays caused by the storms and the wait before the mooring at Saint-Hubert, there still remained fuel for about 14 hours of flight.

The *R-100*'s nose, shaped like an eye, was moored to a socket linked to a hydraulic head the height of which could be adjusted to the ship's position. The responsibility for the operations in Saint-Hubert was delegated to Lieutenant Commander Arthur R. Pressey, leader of an 11 man team, among whom were James H. Temple and George H. Kinch, both trained at Cardington, in October 1928. The visit of the Graf Zeppelin to Lakehurst, New Jersey, had been of benefit to them even if it had provoked many serious logistic problems for the American organizers.

Many visitors, camera at the ready, listened to the radio broadcasts. Never before had the eyes and ears of Canadians been so solicited. After all, it was the first time they had seen a rigid structure airship. Therefore, for a few days at least, the throes of the economic crisis were forgotten.

Numerous radio stations broadcasted the event. The National Railway Company, the CNR, broadcasted the event from sea to sea through its network of stations in Moncton (CNRA), Montreal, Ottawa (CKCH) and Vancouver. CNR Radio had been founded in 1923. In 1924, its affiliated stations were:

Montreal	CKAC (CNRM)
Toronto	CFCA & CKGW (CNRT)
Winnipeg	CKY (CNRW)
Regina	CKCK (CNRR)
Saskatoon	CFQC (CNRS)
Calgary	CFCN & CFAC (CNRC)
Edmonton	CJCA (CNRE)

The event thus brought back to mind aviator Charles Lindbergh's radio broadcast, made in Ottawa, back on July 1, 1927.

Some stations broadcasted through the intermediary of the Canadian Pacific (Railway Company or CPR) which, in 1930, had also decided to explore the field. The CNR had put sleeping cars at the disposal of the journalists; others, as a group, had preferred to rent a house in the area. 20 or so telegraph machines had been installed in a pressroom especially designed for the event. The noise of typewriters soon followed. Many of the journalists stood gaping as they began their account of the airship's arrival. Some of them, positioned in the mooring tower, broadcasted over the radio waves, live. The *R-100* only took about 20 minutes, a record, to get moored to the mooring mast. At 6:30, it was all over.

The first persons to climb on board were the customs officers Arthur Magnan, Arthur Laing, Jean Belise, Paul-E. Jessen, and M. Pilon. Like everyone else, they had to take off their shoes and put on special ones. This was intended to make sure that the friction of the shoes on the floor would not produce sparks. All visitors also had to get rid of their matches. George Meager tells how M. Pressey, the Canadian responsible for the mooring, very moved, greeted the crew members: "Don't mind me chaps; I can't help crying. I'm so pleased to see you." As the aeronauts left the aircraft, more than 10,000 persons had already gathered near the Saint-Hubert tower.

At the outset, visits to the interior were reserved to politicians, military and journalists. The public only had access to the grounds to admire the airship and the mooring tower. The crew of the *R-100* stayed for 12 days in Montreal; they were given rooms in the airport's main terminal.

Many booths and restaurants had been set up on the site to meet the visitors' needs. One could buy soft drinks and juices, as well as many souvenirs. Two small circuses had also come to amuse the visitors. About 30 six-seat planes, from 14 different companies, had been mobilized to offer visitors flights around the *R-100*. As many as 500 persons per day participated in those outings. Pilot Hervé St-Martin made more than anyone (149 in a single day on August 5). Evidently the *R-100* became the principal topic of conversation in trains, tramways, restaurants and offices. On the single day of Sunday, August 3, the *Montreal Daily Star* claimed 300,000 visitors. The entry gates were opened from 10 in the morning until eight in the evening. At night, six spotlights illuminated the dirigible from the ground; mounted on casters they could therefore be easily moved to floodlight the airship.

The Canadian National offered departures from a temporary station on Guy Street, on the hour. A return ticket for adults cost 50 cents and 25 cents for children. The last train left Saint-Hubert for Montreal around 10 p.m.

On August 3, to everyone's dismay, a tear of 1.5 metres was discovered on the top of balloon number 7. It was immediately repaired. Canadian Vickers repaired the fabric of the damaged structure some time before the ship's arrival in Saint-Hubert. From a large storage tank located near the hangar reserved for the journalists, the *R-100* was filled up with hydrogen. The airship managed to make a small 24-hour trip over Ottawa, Kingston, Toronto, Hamilton and Niagara Falls. In Toronto, radio station CFRB broadcast the event live, the reporter Wes McKnight having set himself up on the roof of the building on Jarvis Street.

On August 4th, Major Scott and Commander Colmore gave a press conference at the Montreal Canadian Club. According to them, the mooring tower was the most modern in the world; they were also impressed by the speed of the Saint-Hubert mooring team. On August 5, two women, Miss Leldred Bennett, sister of the Honourable R. B. Bennett, and Lady Perley, wife of Minister George Hasley Perley, put an end to the rumors that women could not visit the *R-100*.

In any event, the two women showed no signs of dizziness or vertigo. At the same time, 13 crewmembers visited Lucerne-en-Québec, a vacation resort in the Laurentians. In the evening, officers Colmore, Booth and Scott were invited to a dinner with the Minister of National Defence, Colonel Ralston.

On August 6, Major Scott and Meteorologist G.B. Giblett were welcomed at the Toronto City Hall. As for Captain Booth and Commander Colmore, they departed for Ottawa where they met with the Governor General of Canada, Mr. Willingdon. Mr. Dennistoun Burney was honored at the Mont-Royal Hotel by the Montreal Advertising Club that presented him with two Majestic Console receivers, provided by Layton Brothers.

The Sun Life Insurance Company had attached to its building, still under construction, an announcement measuring 45 metres long and nine metres high; the biggest sign ever seen in Montreal. It read: "Welcome to Canada *R-100*"; a Union Jack was painted at both ends of the message. The banner, created by Miller Awning, had been attached to hooks at the 22nd floor by employees of Dominion Bridge. It faced Dominion Square and the Windsor Hotel. J.W. Connelle was its designer. Six people from the Brouillette-Stanway Sign Company were required to produce this work.

On Friday, August 8, the crew was hosted by the City of Montreal at the Windsor Hotel. The atmosphere was a little tense due to problems that had arisen between the City of Montreal and the military authorities at the airport. Mr. Bray, Chairman of the city's Executive Committee, had not been allowed to visit the interior of the *R-100*. However, ordinary citizens had been permitted to do so in groups of 10. Frustrated, Mr. Bray recalled the 100 or so policemen of the City of Montreal on duty at Saint-Hubert. The Provincial Police had to take over with fewer than 30 officers. Brigadier General and Commander of Military District no 4, W.B.M King, was forced to call Mr. Bray back and invite him to visit the interior of the *R-100* the very next day. The City of Montreal then re-assigned its policemen to Saint-Hubert, thereby avoiding the need for relief by the soldiers at the military base. At the Windsor Hotel, a surprise awaited 450 guests: Commander Colmore offered his thanks in French first. Scott and Booth also spoke and all thanked the Canadians for their hospitality. Montreal Mayor, Camilien Houde, acted as master of ceremonies. The chef from Windsor Hotel presented to all the guests an enormous cake in the shape of the Saint-Hubert Airport with, obviously, the *R-100* prominently displayed.

Around 18:00, on Sunday, August 10, the captain of the *R-100*, Mr. Booth, from his control cabin with a megaphone, voiced instructions to those responsible for putting away the cables retaining the *R-100*; the ship left the Saint-Hubert Airport around 18:20 under a fine mist, to fly over Ottawa. A large group of people observed the event. A few planes escorted the *R-100* to Montreal. Only one French Canadian was invited on board the *R-100*, and then only at the last minute, in the person of Mr. Corriveau, Deputy Chief Engineer of Canada's Public Works Department. The other guests on that trip over parts of Ontario were:

L.S. Bredner, Squadron Leader
C.P. Edwards, Director of Radio Services, Canadian Navy
J.L. Gordon, Head of Civil Aviation
F.C. Hagdely, Director of the National Film Board
Walter Hose, Navy Chief of Staff
W.A. Laurie, Captain
W.A. Lawrence, Press Agent

George Macdonald, *Canadian Press*
A.G.L. McNaugton, Adjutant, Department of National Defence
J.H. Pudhope, Superintendent, Air Navigation
E.W. Steadman, Chief Air Engineer
W.A. Steel, Major
Frank Tirrell, Official Photographer, National Film Board

The dirigible flew over Parliament Hill a little after 10 p.m. It circled around Ottawa for almost two hours; one radio reporter broadcasted live its every movement. At 4:45 in the morning, on August 11, the *R-100* arrived in Toronto. As it was yet too early to really start admiring the city, the crew headed for Niagara Falls. Along the way the passengers also enjoyed seeing the Welland Canal. The airship then flew over Hamilton, over Lake Ontario and again over Toronto. Wes McKnight, of CFRB, provided live commentary on the visit of the *R-100.*

The flying ship returned to Montreal by way of Cornwall. Upon its arrival in Montreal, it had to hover above the city for more than two hours because of a storm. It moored at Saint-Hubert around 8:15 p.m. The *R-100* surprised everyone when it passed over Montreal a little after four o'clock in the afternoon. Time was taken to fly slowly above the city and particularly over the La Presse Building, on Saint-Jacques Street. To get a better view of the ship, a seventeen-year-old boy, Alphonse Barbeau, climbed onto a hydro pole bearing high tension electrical wires and was electrocuted. One of his friends also got seriously burned. The *R-100 G-FAAV* also flew over Mount Royal to permit a better look at the holy cross erected on its summit. During the trip in Ontario, a portion of the fabric of the *R-100* was torn and one of its engines gave out.

On August 13, in Ottawa, the Canadian Prime Minister, Richard B. Bennett, met Commander Colmore, Major Scott and Dennistoun Burney. The Seagram Company invited the crewmembers to the Mont-Royal Hotel for dinner. In the afternoon, E.L. Johnston, Navigator-in-Chief, made a visit to La Presse in order to thank the newspaper for the publicity afforded the dirigible and its crew. Under a serene sky, the crew left Saint-Hubert around 9:30 p.m., abandoning the mooring tower which would never serve again. On high, from the airship, the passengers could see the headlights of all the cars finding their way back home. They could also hear the last honking of horns dedicated to the *R-100*. The sky shark, as some reporters from La Presse had baptized it, made its way around the island of Montreal and headed up the Saint Lawrence River. About 15 journalists started writing what would likely be their last articles about the *R-100*. Nine other reporters had better luck. Like Jacques-Narcisse Cartier, the reporter from Radio Station CKAC and *La Presse*, they made the return trip of the *R-100* to Cardington. There were representatives from the Southam Press, the *Toronto Evening Telegram*, the *Times of India*, the *Montreal Gazette* (J. Fergus Grant), &c. With 56 individuals on board, the airship departed with 36,000 litres of fuel, 6,3 tons of ballast and 1.5 tons of drinking water.The *R-100* carried little freight upon its return, except for some envelopes especially cancelled for the event, among which was a letter from the Prime Minister of Canada to the British Prime Minister and another one from the Mayor of Montreal, Camilien Houde, to the Lord Mayor of London. There were also two bushels of peaches for his Highness the Prince of Wales, a gift from the St Catharine's Flying Club, according to Meager, or from the farmers of Niagara Falls, according to a reporter from the *Toronto Evening Telegram*, as well as a few peonies for the King and Queen; the flowers had been cut at Rideau Hall by Mrs. Willingdon, wife of Canada's Governor General. The *R-100* left with a repaired fabric, but with one engine less. There was available in Saint-Hubert a new Rolls-Royce engine, but Commander Colmore deemed it useless to spend more time repairing the damage.

The air vessel flew over Quebec at 11:45 p.m., Pointe-au-Père at 2 a.m. and Belle Isle at midday, on August 14. On August 15, the *R-100* encountered a torrential rain that penetrated inside the dirigible, forcing the crew to stop using the electric stove. There were a few power failures. Even the reporters had to pump fuel from the tanks to the engines. The water was stored as ballast but, quickly, the reservoirs were filled to capacity.

During its return, the *R-100* flew over the steamers Empress of Australia and SS Bever Brae. Radio communication was maintained with the Saint-Hubert base for the first 960 kilometres; then with the stations at Louisbourg, Portishead, Somerset and finally with Cardington for the last 587 kilometres.

A wrong move on the part of a young and inexperienced elevation plane operator put the crew in danger. Captain Booth himself had to intervene in order to raise the ship, which was only 150 metres above the sea, to a higher altitude. Upon its arrival at Bristol, in Great Britain, two planes escorted the *R-100*.

The return to Great Britain took 57 hours and 36 minutes, the winds blowing from west to east. 12,110 litres of fuel remained in the tanks. The only time that the *R-100* made a dive so steep as to make a pen roll off a table was during its mooring at the Cardington tower, on August 16.

After a 19 day absence, no crowd comparable to that of Saint-Hubert gathered to greet the arrival of the 56 passengers. However, the Secretary of State for Air, Lord Thomson of Cardington, was present. The *R-100*, like all the airships that had crossed the Atlantic, had accomplished a feat; moreover, it had broken all previous time records for the return trip.

In his excellent book, *Airship Navigator*, E.A. Johnston, the son of officer Ernest Johnston, tells of certain crewmembers monitoring the *R-100* spending most of their time drinking rum and the plundering of several cabins. Officer Ernest Johnston apparently punched an individual who was interfering with him. Officer Atherstone indicated to Johnston that he understood his reaction but that it was inappropriate. Those accused were suspended by Scott.

As of August 17, the *R-100* was housed in its hangar. Despite a mission crowned with success, it was deflated and its keel inspected, never again to be taken out of its hangar except to be demolished. The *R-100* thus died while at rest. It was sold later to Elton, Levy & Co. for the value of the metal. The engines and the balloons however, were salvaged. The rest was destroyed in December, 1931. Charles Dennistoun Burney, the original developer of the *R-100* then turned his efforts to building automobiles that bore his name.

The *R-100* had made 10 flights for a total of 276 hours; it had stayed moored for 297 hours. The cost of the dirigible was estimated at £470,000.

From then on, all energy and attention were directed towards the development of the *R-101* and to its trip to Karachi, now part of Pakistan. It was first divided in half at Ring nº 8, to enlarge it by 13.7 metres.

The *R-100* flies close to the belfry of the Saint Francis Church, île d'Orleans.
Louis Emard Collection.

The *R-100* above the St. Lawrence River.
Library and Archives Canada Collection, PA 119889. Original copyright Interprovincial Airways Ltd.

R-100 reaches Quebec, near the Château Frontenac Hotel.
Postcard published by W.B. Edwards. Yves Beauregard Collection.

The *R-100* above the St. Lawrence River.
Library and Archives Canada Collection, PA 119889. Original copyright Interprovincial Airways Ltd.

At nightfall, the arrival of the *R-100*. Like a fish circling the hook, the airship is about to be moored. Note the lights of the tower.

Photo taken at 5 h 15 a.m., on August 1st, 1930, upon the *R-100*'s arrival at Saint-Hubert.
Société historique du Marigot Collection.

Mooring the nose of *R-100* to the mast of the tower. The passengers' door is open.
Library and Archives Canada Collection. PA 117709.

View of the landing platform of the *R-100*.

Louis Emard Collection.

Among the first to congratulate the officers of the *R-100* was the Mayor of Montreal, Camillien Houde, here in the center, cane in hand.
Library and Archives Canada Collection. PA 117677.

Several important personalities, a little after the arrival of the *R-100*: Mayor of Montreal, Camillien Houde, Sir Charles Dennistoun Burney, Director of the Airship Guarantee Company, Lieutenant-colonel Reginald Colmore, Director of Political planning for British airships, Colonel James L. Ralston, Canadian Defence Minister, Major George H. Scott, Captain George F. Meager and Commander Ralph S. Booth.

The «builder » of the *R-100*, Sir Dennistoun Burney, the Canadian Defence Minister, J. L. Ralston, Major G. H. Scott and R. B. Colmore.
Library and Archives Canada Collection, PA 117696.

Marc-André Valiquette Collection.

« Yvette ». *Michel Pagé Collection.*

Marie-Paule Shaffer Levac Collection.

Library and Archives Canada Collection, PA 117698.

Photo by B.J. Hebert, Saint-Hyacinthe.
Simon Beauregard Collection.

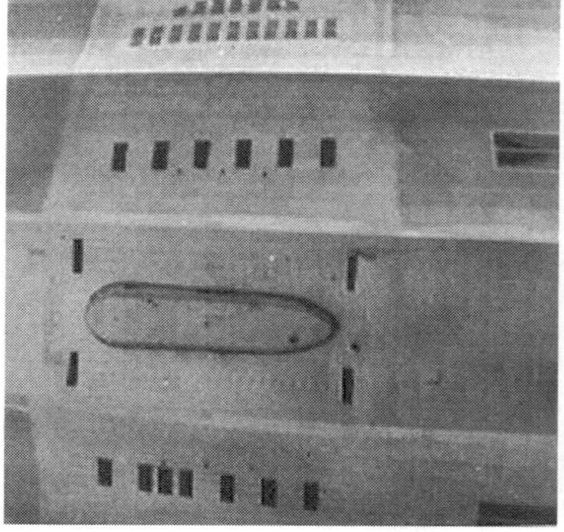

Observe the numerous cables at the back of the *R-100*.
Photo Rice. Simon Beauregard Collection.

Nelson Newberger came to visit the *R-100* by motorcycle.
McCord Museum of Canadian History Collection, Montreal, n° 73036033.

Serge Juneau Collection .

Serge Juneau Collection .

Raymond Plante Collection.

One of many photos of people who immortalized themselves on film in front of the *R-100*.
Société historique et culturelle du Marigot Collection.

A fine touch from the photographer who managed to get almost the entire crowd to pose.
Photo La Presse. Jean-Luc Allard Collection.

Royal Canadian Air Force Collection .

Judging from the flying neckties, it is easy to imagine just how windy it was in Saint-Hubert on that day.
Société historique et culturelle du Marigot Collection.

This phtograph was used to create a poster.
Marc-André Valiquette Collection.

Some cars were able to park near the airship, giving their occupants an exceptional view of the *R-100*.
Société historique et culturelle du Marigot Collection.

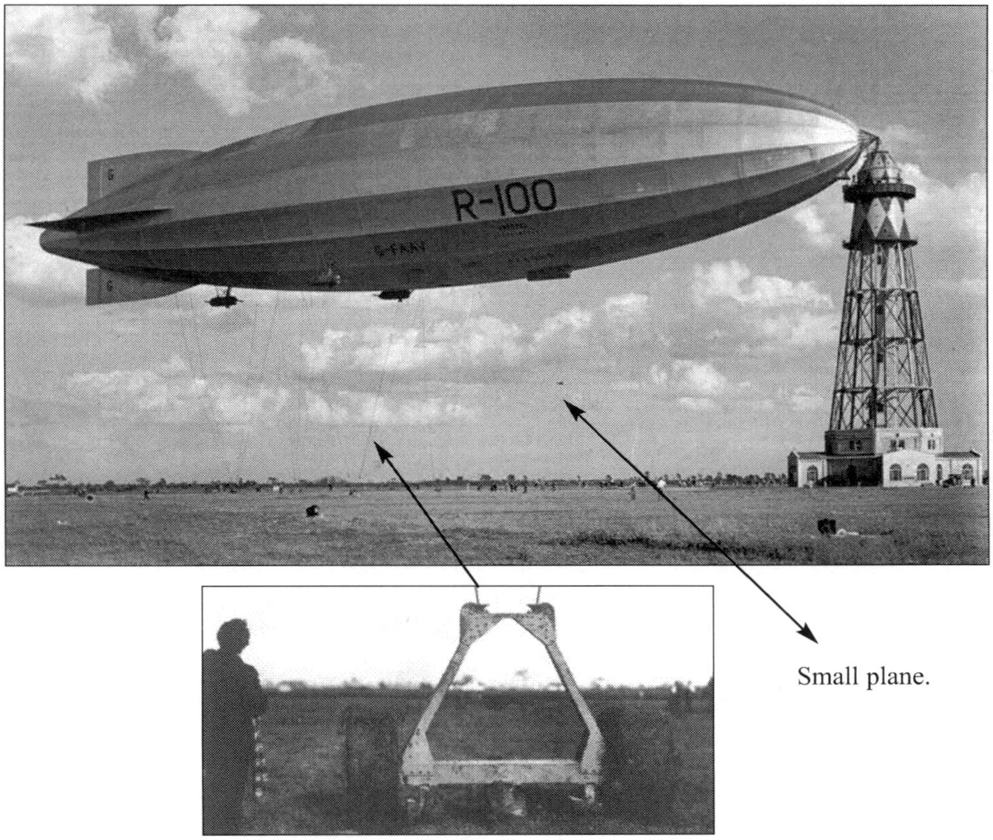

The two cables stretched towards the top are from the airship *R-100*.
Library and Archives Canada Collection, PA 117704.

View of the fence surrounding the site.

Under the *R-100*, one can see the church at Saint-Hubert.
Airship Heritage Trust Collection.

The photo that best depicts the location of the damage on the fabric.

View of the Saint-Hubert Airport. Note the many white tents, on the left, as well as a bus at the edge of the lane.
Library and Archives Canada Collection, PA122173.

R-100
G-FAAV

R-100
G-FAAV

Montreal policemen on duty, August 7, 1930. On the left, the silhouette of *R-100.*
Musée de la police de l'île de Montréal Collection.

During first days following the arrival of *R-100*, there were frequent excursions for visitors in these kinds of planes.
Louis Emard Collection.

Motorcycles were popular, especially among policemen.
Louis Emard Collection.

Jean Mercure Collection.

Louis Emard Collection .

Mid Week Pictorial.

Repairing part of the fabric during which a few pieces, thrown on the ground, made some collectors happy.
Chandler Photograph.

Repairing the fabric in the Vickers workshops, in Montreal, on August 2nd, 1930.
Library and Archives Canada Collection, C 83934.

The *R-100* was illuminated by six powerful spotlights.
Library and Archives Canada Collection, PA 119891.

The *R-100* by day. The building in the foreground was the assignment center for journalists; it was also used for official receptions. Note that a plane has been moved, on the left, for the occasion. A half-dozen photographers, close to the fence, seem to appreciate their viewing angle of the *R-100*.

Mid Week Pictorial.

Three means of transport: the car, the airship and the plane. In the background, Mount Saint-Bruno.
Library and Archives Canada Collection, PA 117710.

At the extreme-right-hand side, the railway line of the CN. On the left, note the two sidings especially arranged for passengers wishing to visit the *R-100*.
Canadian National Collection.

Note the very official and formal attitude adopted for visiting the *R-100* and in particular the white shirts systematically worn by all the men.
Marc-André Valiquette Collection.

Marcel Paquette Collection.

Passengers disembarking from the Canadian National train. They have the opportunity to see a plane taking off and also to admire the *R-100*.
Canadian National Collection.

Overall view. In the background, the very distinctive Mount Saint-Bruno.

Some passengers have the privilege of observing the landing platform and observation deck of the *R-100* more closely.

Crowd gathered on Mount-Royal to watch the *R-100* cross the skyline of Montreal.
McCord Museum of Canadian History Collection, Montreal, nº 025-1054.

The *R-100* above Montreal. In the background, the new du Havre Bridge which became, in 1934, the Jacques-Cartier Bridge.
Christian Paquin Collection.

R-100 passing near the Sun Life Building, still under construction and bearing the banner « Welcome *R-100* to Canada »
Christian Paquin Collection.

Near Dominion Square, in Montreal.

Near Mount Royal, in Montreal

Above the Windsor Station of the Canadian Pacific Railway, in Montreal.
Photo Harry Sutcliffe nº 21213.

Many people took to the roofs of buildings, some in extremely risky positions.
Société historique du Marigot Collection..

The airship *R-100* flying over Mount Royal, in Montreal.

Yvon Ménard Collection.

Above Montreal. *Christian Paquin Collection.*

Inside the command cabin of the *R-100.*

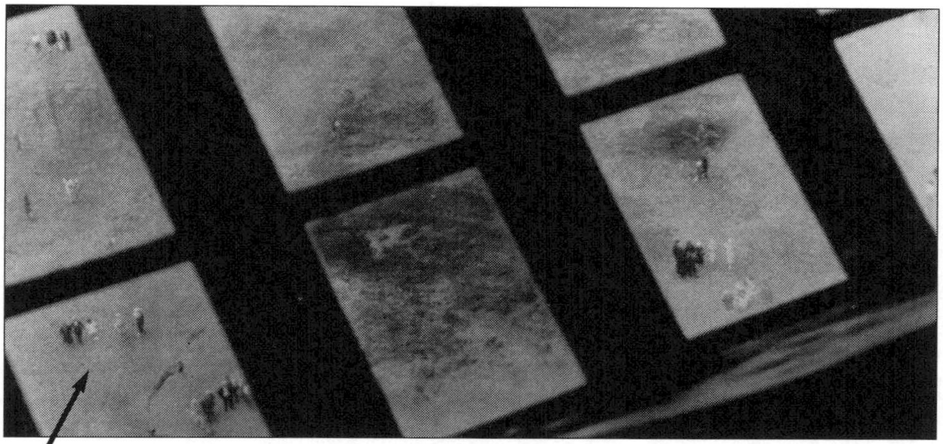

Observation gallery.
Library and Archives Canada Collection, PA 149475.

Serge Juneau Collection.

Cockpit.
Library and Archives Canada Collection, PA 125414.

A few journalists in the observation deck of the airship during the trip to Ontario.The first person leaning over is George Macdonald, of the *Canadian Press*.

Above the Parliament in Ottawa.
Library and Archives Canada Collection, PA 207267.

Above the Canada Life Building under construction on University Avenue in Toronto.

Above Toronto.
Marc-André Valiquette Collection.

Back from its trip to Ontario, escorted by a small plane, the *R-100* flies over the Lachine Canal.
Mid-Week Pictorial, August 29, 1929.

It is hard to believe that the airship managed to fly so close to the waters of Lake St. Louis.
Marc-André Valiquette Collection.

Goodbye Montreal, Longueuil...
Marc-André Valiquette Collection.

Quebec Bridge.
Yves Beauregard Collection.

The *R-100*, back from its trip to Canada, approaching the mooring tower at Cardington.
L'Illustration nº 4564.

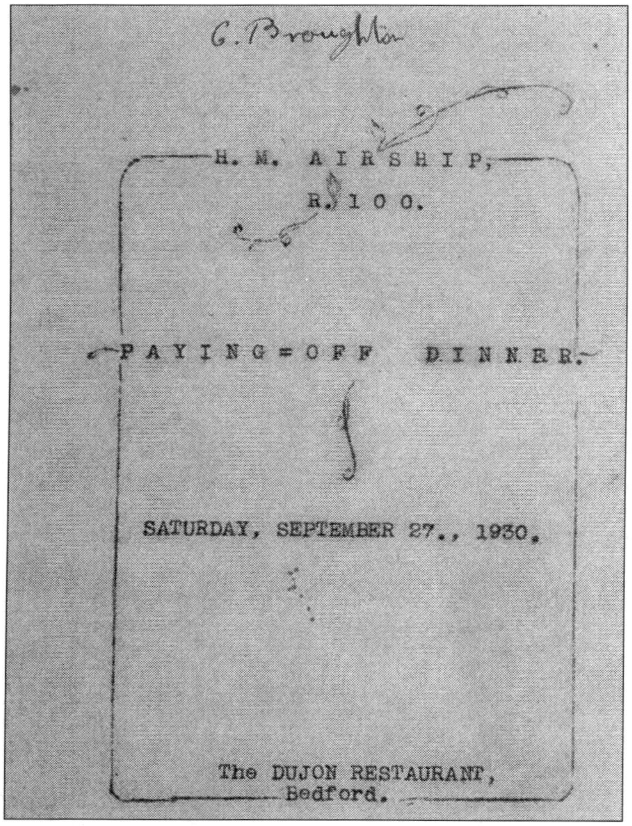
G. Broughton

H. M. AIRSHIP,
R. 100.

PAYING=OFF DINNER.

SATURDAY, SEPTEMBER 27., 1930.

The DUJON RESTAURANT,
Bedford.

On September 27, 1930, the crewmembers of the *R-100* were received for dinner at the Dujon Restaurant, in Bedford. They autographed the menus as they had done during the reception in Montreal.
Nigel Caley Collection.

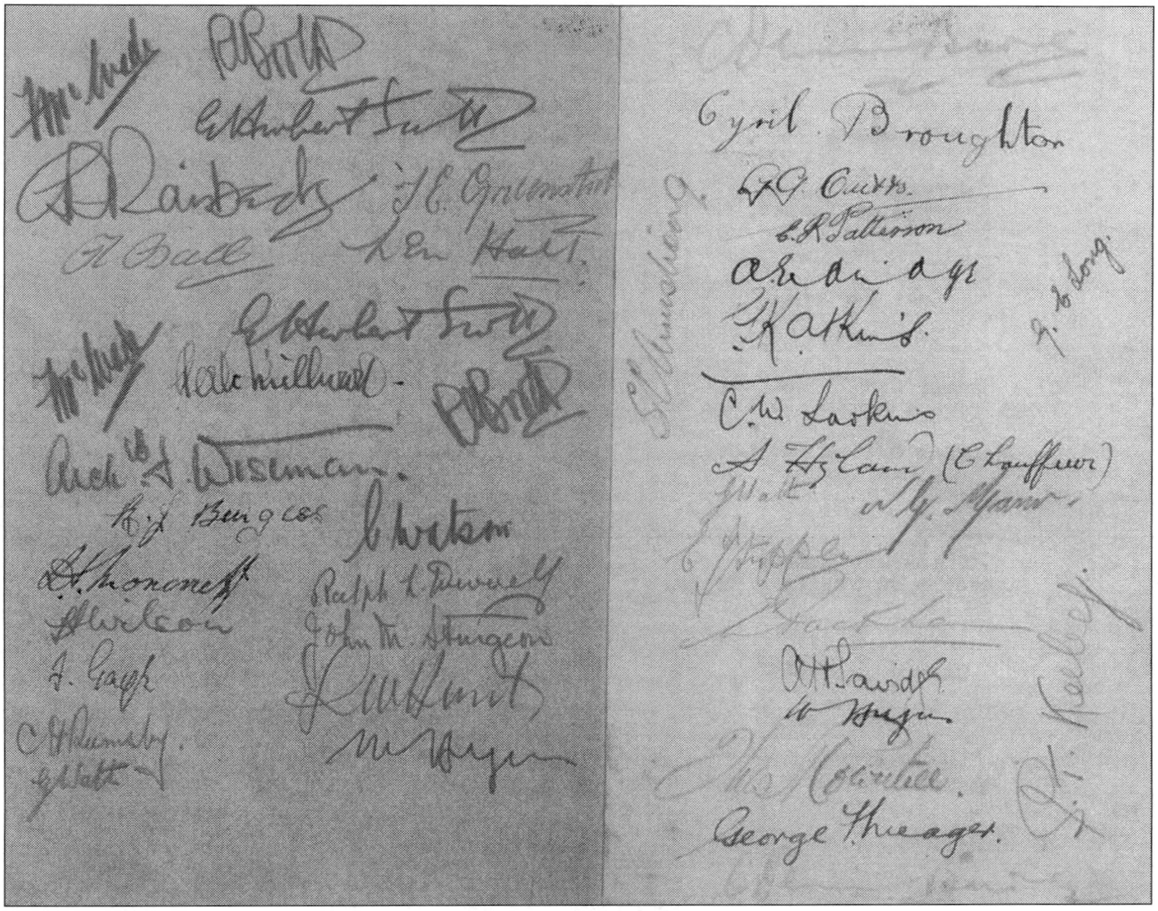

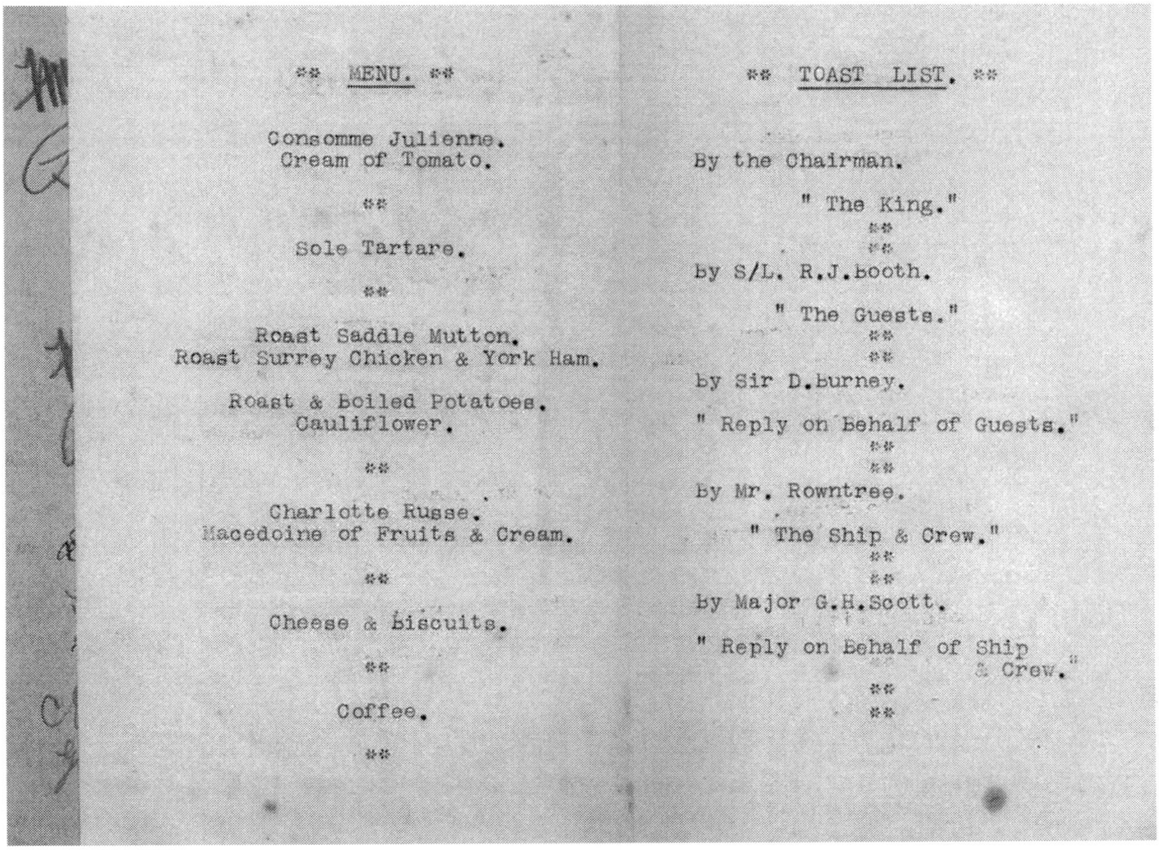

** MENU. **

Consomme Julienne.
Cream of Tomato.

**

Sole Tartare.

**

Roast Saddle Mutton.
Roast Surrey Chicken & York Ham.

Roast & Boiled Potatoes.
Cauliflower.

**

Charlotte Russe.
Macedoine of Fruits & Cream.

**

Cheese & Biscuits.

**

Coffee.

**

** TOAST LIST. **

By the Chairman.

" The King."

**
**

By S/L. R.J.Booth.

" The Guests."

**
**

By Sir D.Burney.

" Reply on Behalf of Guests."

**
**

By Mr. Rowntree.

" The Ship & Crew."

**
**

By Major G.H.Scott.

" Reply on Behalf of Ship & Crew."

**
**

January 1931: dismantling the *R-100*.
Airship Heritage Trust Collection.

For a long time, enthousiasts of the *R-100* had been amazed by the simultaneous passing of the boat under the Jacques-Cartier Bridge while the *R-100* passed overhead but the image was only a photomontage by Harry Sutcliffe. Below, a less subtle montage.
Marc-André Valiquette Collection.

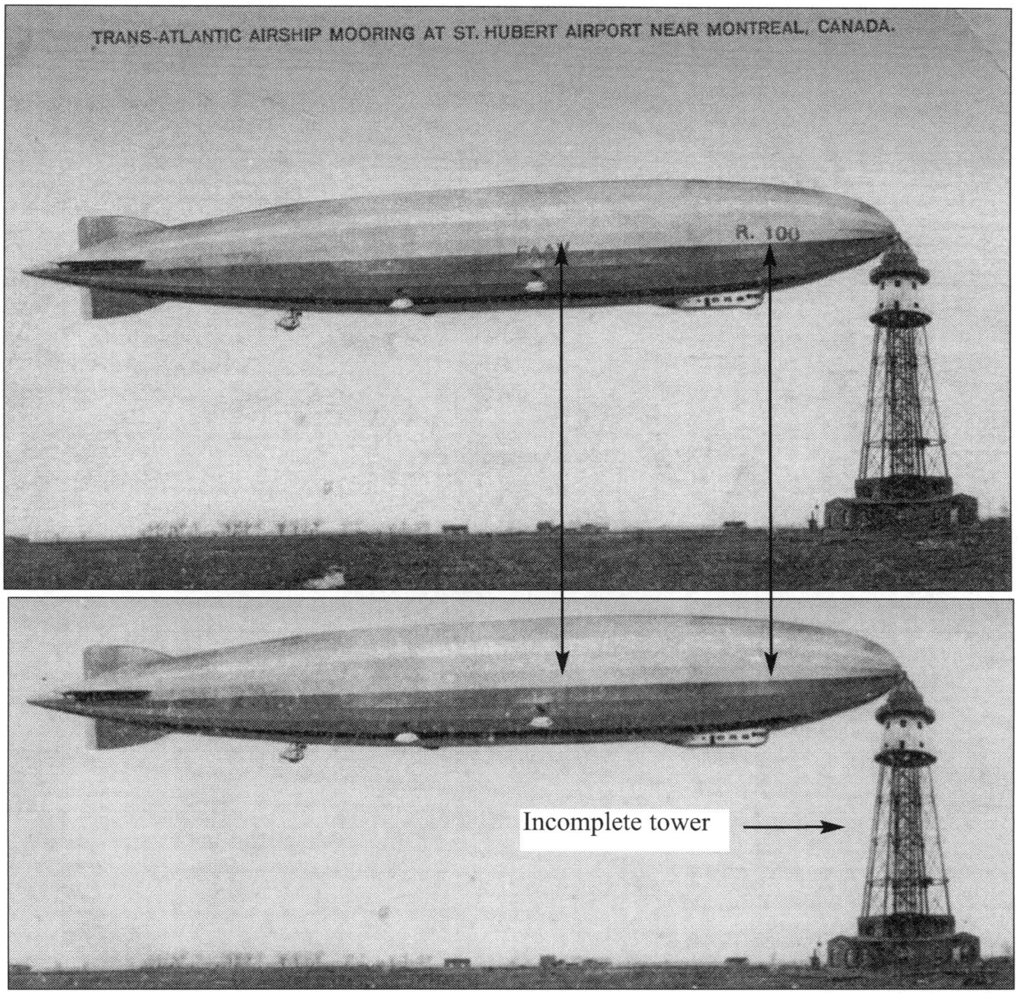

Page from *La Presse,* May 4, 1929, announcing the coming of the *R-100* to Saint-Hubert.
Société historique et culturelle du Marigot Collection.

Postcard published in the bulletin of the Club des cartophiles québécois, Postcards, 1996,
Cartes postales, nº 20 1996.

Postcard sent, on August 5, 1930, from the Saint-Hubert Airport. The sender writes : “ Busy on *R-100* assignment. ”
Société historique et culturelle du Marigot Collection.

WELCOME R-100
R-100
Of Course "NESTLÉ'S"
NESTLÉ

R-100
R-100
carried
Waterman's

BURNETT'S
a ça de
mieux
SIR ROBERT
BURNETT'S
LONDON DRY GIN
26.6 onces $2.95 – 40 onces $4.35
ETABLIE EN 1770

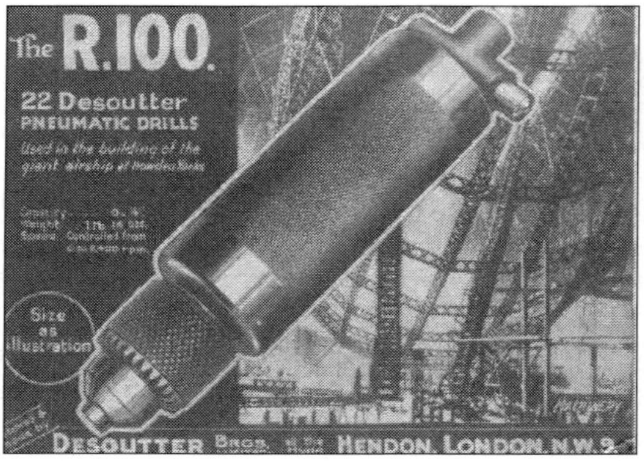
The R.100.
22 Desoutter
PNEUMATIC DRILLS
Size as illustration
DESOUTTER BROS. HENDON, LONDON, N.W.9.

Supreme
Save the
POKER
HANDS
Turret
CIGARETT
TURR
Cigarettes

LEURS PREMIÈRES CIGARETTES
FURENT DES WINCHESTER
Cigarettes
Winchester
D'un Mélange Parfait

Arthur Lapierre interpreted a song, in 1930, composed by Romeo Beaudry and entitled *C'est l' R-100.*
(*It's the R-100*)

Harold Leonard, of the Windsor Hotel Orchestra, and Glenn Adney composed a song which was interpreted during the stay of *R-100*: The *R-100*, published by Popular Music Publishing.

Soucy, Lafleur and Montpetit interpreted a song: *Salut au R-100* (*Salute to the R-100*). The song was published under Starr label at the end of August, 1930.

Piece of a puzzle sold for only 15 cents.
Marc-André Valiquette Collection.

Commemorative flags of the *R-100*.
Société historique et culturelle du Marigot Collection.

Plastic model of the *R-100*.

Marc-André Valiquette Collection.

Toujours l'*R-100*

from Mary Travers (la Bolduc) Paroles and musique : Mary Travers « La Bolduc », 1930.

M'as t'changer d'nom mon Jean
Pis m'as t'app'ler l'R-100
Ti-Rouge l'R-100, Ti-Gus l'R-100, Ti-Pit l'R-100
Moi j'trouve ça du bon sens
C'est les culottes l'R-100, les pyjamas l'R-100
Brassières l'R-100, jarretières l'R-100
Tout l'monde parle de l'R-100.

Mais voilà qu'tu vois pu clair tes culottes sont d'vant derrière
Quand même qu'on est habitants faut montrer qu'on a du bons sens
Va donc mettre ton Prince Albert pour aller à Saint-Hubert
On partira le coeur content pareil comme deux jeunes amants.

Ma belle-mère qui est en arrière avec sa vieille tabatière
R'garde-moi donc son jupon blanc a va l'perdre en s'en allant
Mais rendue à Saint-Hubert voilà son jupon par terre
C'est le garçon de Vincent qui l'a pris pour s'moucher d'dans.

C'est la vieille fille engagère avec les yeux virés à l'envers
Ça faisait presque un an qu'a t'nait ses gaz pour l'R-100
Un ami de Saint-Hubert avec son nez r'troussé en l'air
Il dit à son ami Jean : « Tu trouves pas qu'ça sent l'hareng ? »

Ma belle-soeur à Saint-Hubert qui fortillait comme un ver
Tout d'un coup a voit l'R-100 voilà qu'a veut sauter d'dans
Quand elle a vu mon beau-frère la voilà les pattes en l'air
Son corset volait au vent, je pensais qu'c'était l'R-100.

CHEMIN DE FER CANADIEN NATIONAL

Le R-100 à St-Hubert

Durant la période de visite

LES TRAINS QUITTERONT LA
GARE TEMPORAIRE DE LA RUE GUY
A TOUTES LES HEURES

De 10.00 a.m. à 6 p.m.
Heure d'été

DERNIER TRAIN DE MONTREAL (RUE GUY)
6 p.m. Heure d'été

Montréal à Saint-Hubert (aller et retour) Adultes **50c** Enfants **25c**

AU RETOUR, des trains spéciaux quitteront l'aérodrome de Saint-Hubert à des heures commodes. Dernier train à 10 p.m. heure d'été

BILLETS ALLER ET RETOUR EN VENTE AUX ENDROITS SUIVANTS: Gares temporaires du chemin de fer Canadien National, rues Guy et Sebastopol, aussi au bureau des billets de la ville, 384 rue Saint-Jacques; gare Bonaventure; terminus du tunnel; gare de la rue S.-Catherine est; gare Saint-Henri; gare Pointe-Saint-Charles; terminus des tramways, 121 Craig ouest; hôtels Mont-Royal, Windsor et Queens; James-A. Ogilvy Ltd; T. Eaton Co. Ltd; Dupuis Frères Ltée; Le Devoir, 480 Notre-Dame est; Les Voyages Hone, immeuble University Tower.

CANADIEN NATIONAL

Publicity in 1930.

Souvenir of the 60th anniversary of the arrival of *R-100* in Canada.
National Aviation Museum.
Louis Emard Collection.

Pin.
Louis Emard Collection.

More pins.

Advertisement in 1930. The bottle is used as mooring mast.

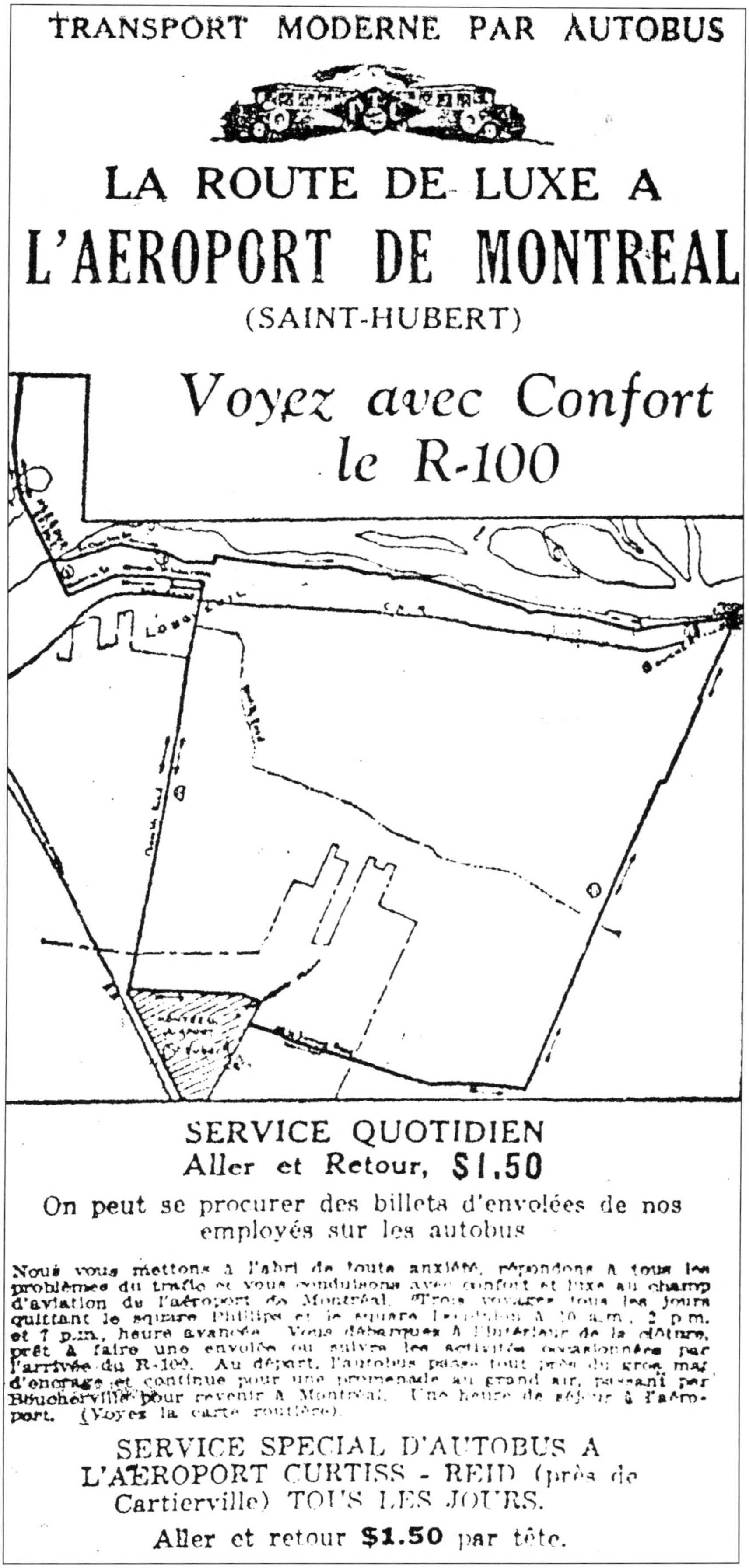

Advertisement in 1930.

Airship *R-101*, with its outer cover on, is almost ready to leave its hangar. Note the dog at the bottom of the photograph.
Barratt Collection.

Characteristics of the *R–101*

The *R-101* was designed by Lieutenant-Colonel Vincent Richmond, one of the designers of the *R-38* that had split in two in 1921, causing the death of 44 people. Michael Rope also played an eminent role in the design of the *R-101*. The function of Chief Calculator, the equivalent of Nevil Shute's position in the *R-100*, had been given to Harold Roxbee Cox.

The pisciform structure of the *R-101* was shaped fish-like by 18 transversal rings, of a polygonal form, joined to 15 main longitudinal girders and to 15 intermediary ones, all made of stainless tubular steel. These girders were triangular. The elements of the structure, designed by the engineers Boulton & Paul, were constructed at Norwich, and then assembled at Cardington. The chief designer of Boulton & Paul, John Dudley North, also contributed to the design of the pieces.

The magazine *La nature* explains the *R-101*'s structure:
"All its girders pass along the outside of the coupling rings thus leaving the dirigible's inner space entirely free. The rings are polygon in shape and are built, like the longitudinal girders, of triangular elements with tubular buttresses made of high tensile stainless steel. The rings, except those on the extremities, had no radial bracing, thus leaving all the room necessary inside the keel for communicating the length of the ship and for storing the fuel and the ballast. As well, absolutely no rivets were used in the assembly, all of which was put together with bolts: this also explains the relative ease with which the dirigible's body had been elongated, by adding a certain number of coupling rings and girders."

The dirigible had 17 gasbags capable of storing 156,000 cubic metres of hydrogen required to provide the lift necessary for the airship and enable it to raise 167.2 tons of material. The fabric coating of the balloons was made of cotton to which goldbeater's skin was glued. Those skins, imported from Chicago, in the United States, as noted earlier, came from the intestines of oxen. The coating of the *R-101* was quite similar to that of the *R-100*. To ensure it remained impermeable, varnish was applied to both sides of the fabric and a layer of beeswax and of aluminum powder was added on the exterior. The envelope was silvery on the outside, but red on the inside, to protect it from ultraviolet rays.

The four-bladed wood propellers were activated by Beardmore Mark 1 Tornado diesel engines with eight in-line cylinders furnishing 585 h.p. at 900 rotations per minute. The most remote engine, near the tail, was located on the airship's centre line; the other engines were grouped in pairs, two in the front and two at the back. The engines were cooled by the evaporation of water passing through radiators. Diesel oil was chosen because it was less expensive: £5 per ton compared to £27 per ton for gasoline. Plus, diesel was thought to be safer in the tropics as it is less volatile than gasoline. A tank was located inside the airship. The same tank was connected by gravity to another transfer tank. From there, the fuel could be conveyed by compressed air to any tank above the engines, or could be used as ballast in case of an emergency. On top of each of the engines there were two fuel tanks of 850 litres each. Overall, the *R-101*c could count on 29 tons of fuel. A supervising mechanic, appointed to each cabin, received his instructions by telegraph.

The ballast was composed of water stored in tanks of half a ton each. The total weight of the ballast was eight tons.

The *R-101*a and the *R-101*b only had four forward thrust or "cruise" mode engines; it was not until the appearance of the elongated version, the *R-101*c, that five engines in "cruise" mode, two of which could function in reverse, could be counted on. In the first versions the problem was caused by reserving one engine to provide reverse thrust. It was completely useless in other circumstances. When the ship was leaving the mooring tower, the four forward thrust engines had to be turned off, using only the one that functioned in reverse mode.

Strange as it may seem, the *R-101* had a smoking-room! The floor of that room was made of metal in order to decrease fire hazards. The passengers also had a 9 by 18 metre living room they could relax in it, while playing certain social games such as bridge. The *R-101* was also endowed with promenade decks on each side and therefore, had much better lighting than the *R-100*. Very luxurious for an airship, the *R-101* was compared to a hotel floating in the air.

The *R-101*a did not respect the original contract: it managed to lift a bulk load of only 148.6 tons instead of 150 and a payload of 35 tons instead of 60. Moreover, the airship ship weighed 113.6 tons instead of the planned 90 tons. It became evident that the structure had to be lightened and something had to be done to prevent the loss of hydrogen from the small balloons or gasbags found to be too porous.

The *R-101* cost £562,000, much more than had been foreseen, and its construction lasted six years. Many tests were made on models, between 1924 and 1926, at the National Physical Laboratory, before assembly could begin in Cardington.

THE TESTS

The airship was ready on October 8, 1929, but it would take four more days before some 400 persons, hauling on cables, managed to get the *R-101*a out of the hangar at Cardington.

Around 5:45, the nose of the airship came out first. Captain Scott led the operations on the ground. The entire ship was pulled out in less than three minutes and brought close to the mooring tower. Ballast was then jettisoned to raise the *R-101*a to 183 metres, to be brought back to about 15 metres from the mooring point. Control for the rest of the operation was turned over to the director of the mooring tower.

On October 14, 1929, a little before noon, the airship made its first flight, lasting five hours and 33 minutes, above Bedford and London. It was an important experience as none of the crewmembers had flown in a dirigible for at least three years, since the era of the R-33. Lord Thomson of Cardington had only come back to power since the month of June, 1929; his absence having prevailed during the entire time the airships *R-100* and *R-101* were developed. Still, the *R-101*, the airship of State, owed its success to him. He insisted on taking part in the second test and even submitted a list of passengers.

Four days later, his wish was granted as he was taken on board the airship for a test of nine hours and 38 minutes. The dirigible having departed a little after eight in the morning, flew over Birmingham and Nottingham. While disembarking, Thomson put a lot of pressure on the *R-101*a construction team by declaring that he was expecting the ship to be ready for a trip by Christmas. Thomson knew quite well that goal could not be met. Later, he qualified his proposition insisting on the need and the priority of respecting a safe development program. Had he been influenced by the writings of Dennistoun Burney, who doubted very much that airships could be an interesting commercial investment? On November 1st, early in the morning, the *R-101* was taken out of its hangar and, a little before 10 o'clock, it left the Cardington mooring mast for a seven hour and 15 minute flight, mostly in the Midlands region. Samuel Hoare, former Air Minister under the Conservative government, took part on this third flight.

The following evening, another test flight of 14 hours and two minutes was made around the Isle of Wight. That was the first time the *R-101*a had flown by night. Upon its return, the airship hit the top of the tower and its nose lost one "tooth". The incident was minor and the damage immediately repaired. Nonetheless, the officers started to object to the ever-growing number of guests. Instead of being focused on the tests, the operation was becoming primarily a matter of public relations.

On November 8, the *R-101* was limited to flying for only three hours and four minutes in the vicinity of Bedford. On November 14, during its sixth test, it made a flight much similar to that of the 8th of November. The officers' complaints were in vain because the number of passengers reached 32, including the Director of Civil Aviation, Sefton Brancker, and 10 members of Parliament.

During November 17 and 18, the *R-101* flew for 30 hours and 41 minutes, mostly above Newcastle and Edinburgh. That was the real endurance test for the airship. Moreover, there were no invited guests, only individuals strictly involved in the development of the dirigible.

The *R-101*a, even when it was left moored, still experienced problems when it rained and the wind blew. Once, when the machine was being placed inside the hangar, 15 pierced hydrogen bags were discovered. On November 30, the *R-101* was again put in its hangar to undergo some modifications. More than 4 000 stickers were used to cover the holes in the bags containing the hydrogen. Considered useless, the power steering unit was abandoned along with the stays retaining the porous gasbags; a reduction in the weight of the ship was thereby achieved but at the cost of creating a chronic instability. In fact, no longer tightly enclosed, the gasbags could move about freely to a distance reaching at times four metres. Worse, they could rub up against the upper longitudinal beam and rip their envelope, causing a loss of hydrogen.

As the power to lift had to be considerably increased it was decided to insert another small hydrogen balloon. Accordingly, the *R-101* had to be cut in two, at its centre and lengthened. Lord Thomson of Cardington gave authorization on December 7, 1929. Also, the small observation platform on the upper forward part of the airship was removed. In June, the *R-101*b was prepared for an air show with the Royal Air Force, at Hendon, at which King George V and Queen Mary were to attend. But during the mooring a 43 metre long tear appeared on the envelope. The next day, another smaller tear appeared on the roof of the airship.

On Thursday, June 26, a local test flight of about four hours and 25 minutes was carried out. The next day, the *R-101*b flew for 12 hours and 33 minutes. Finally, as planned, the *R-101*b went to the pageant on Saturday. The *R-101* being used however, was not yet the elongated one. Captain Meager replaced Atherestone, the First Officer, who was travelling to Canada. Captain Meager's account of that flight underlined the dangers that awaited *R-101*b. During the first hours there were no problems, but as time passed, the airship started losing altitude and became heavy, particularly after 10 hours of flight. Water ballast was jettisoned to prevent a crash.

Captain Meager warned his superiors of the need to make major corrections before another flight, otherwise grave danger awaited the airship. A definite hydrogen leak from the small gas balloons was the root problem. The same one, it seems, that caused the accident at Beauvais after a seven hour and 30 minute flight under difficult weather conditions. Captain Meager deemed himself lucky to have returned unharmed from that trip.

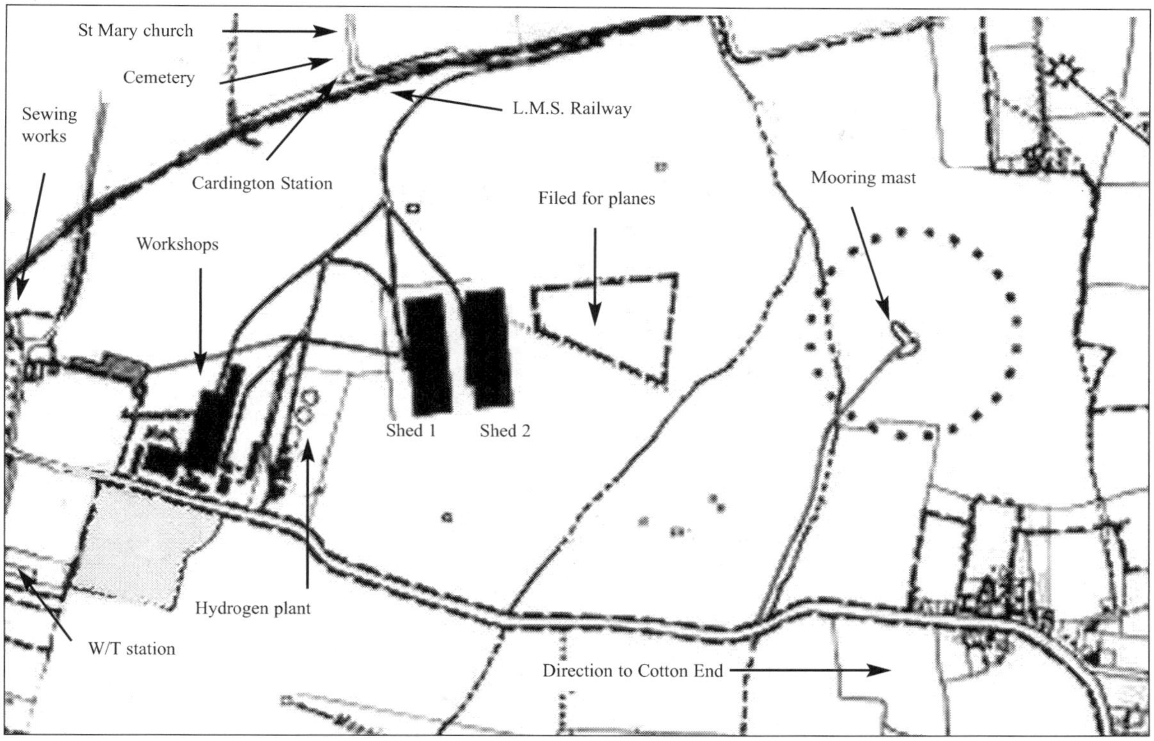

Cardington airship installations, in 1930.

Construction of the *R-101*. Photo below: fuel tank.

Pod for the engines of the *R-101*.

Workers of the fabric shop at Cardington cleaning the oxen intestines, used in the manufacture of gasbags for the *R-101*.
Airship Heritage Trust Collection.

Worker on the longitudinal corridor, inside the *R-101*.
Airship Heritage Trust Collection.

Manufacture of the first ring for the *R-101*. The presence of a person, in the lower righthand corner, gives us an idea of the vastness of the factory.
Airship Heritage Trust Collection.

Partially inflated gasbag.

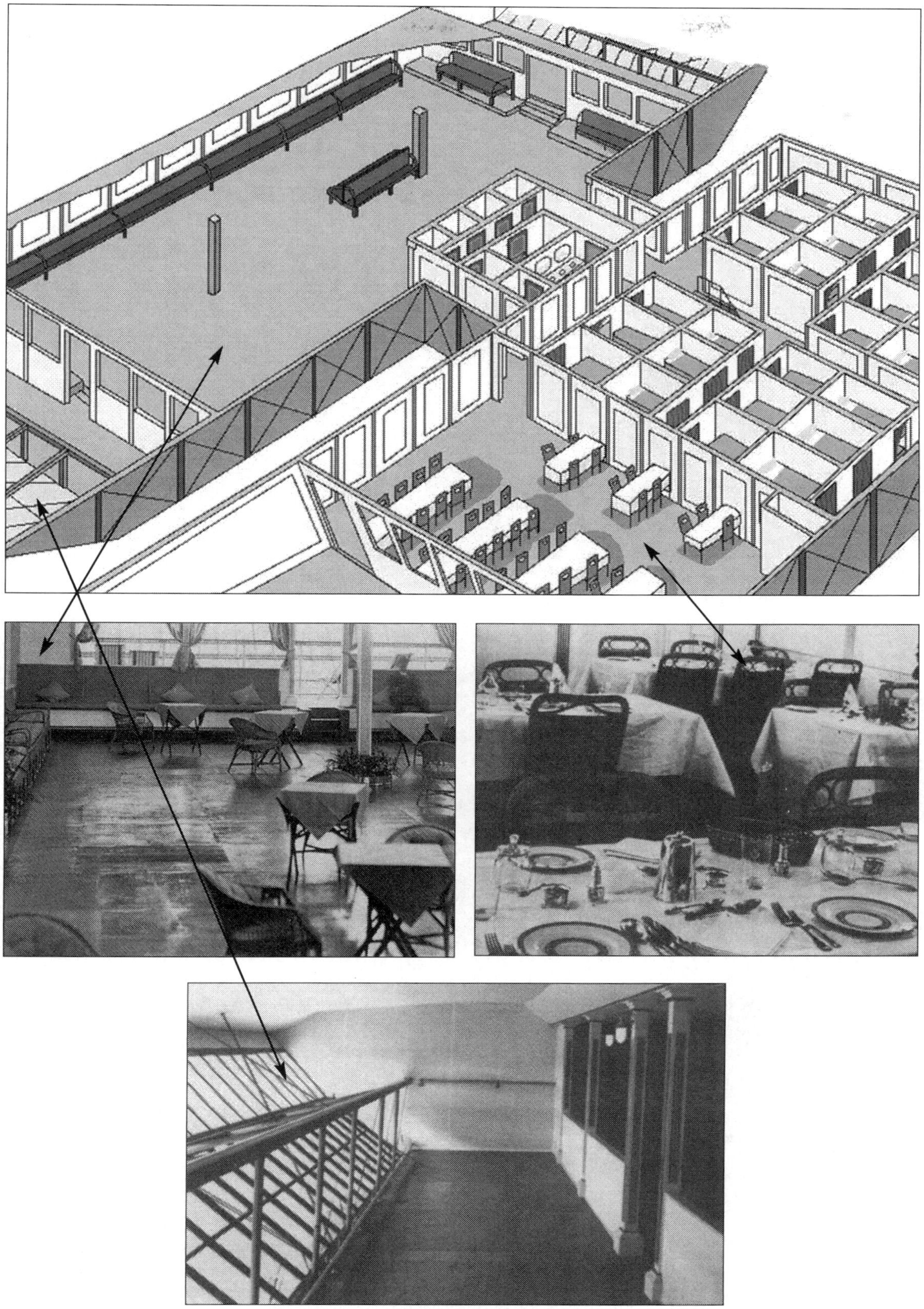

Adapted plan of the *London Illustrated News*. Airship Heritage Trust Collection. Left corner: the Smoking Room; right corner: the Dining Salon; in the center: the Promenade Deck, *Crown copyrigh*t. *http://www.airshipsonline.com*

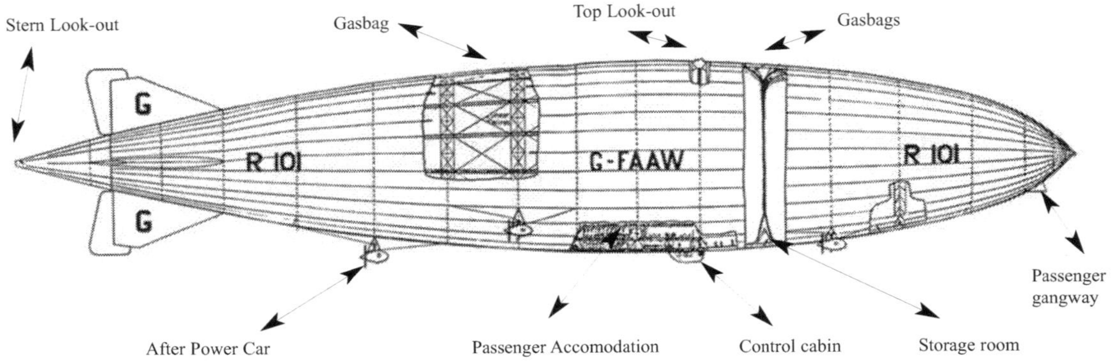

Airship Heritage Trust Collection.

The *R-101* moored to the Cardington tower.
Société historique et culturelle du Marigot Collection

Airship Heritage Trust Collection
http://www.airshipsonline.com

The *R-101* just out of its hangar.

R-101 tries to restore its equilibrium by releasing some of its ballast.
Photograph at bottom: the airship well moored to its anchoring tower.
L'Illustration n° 4520.

Photo from the Cardington mooring tower.

Superb photograph of the elegant *R-101* above the western part of London, October 14, 1929.
Aerofilms.

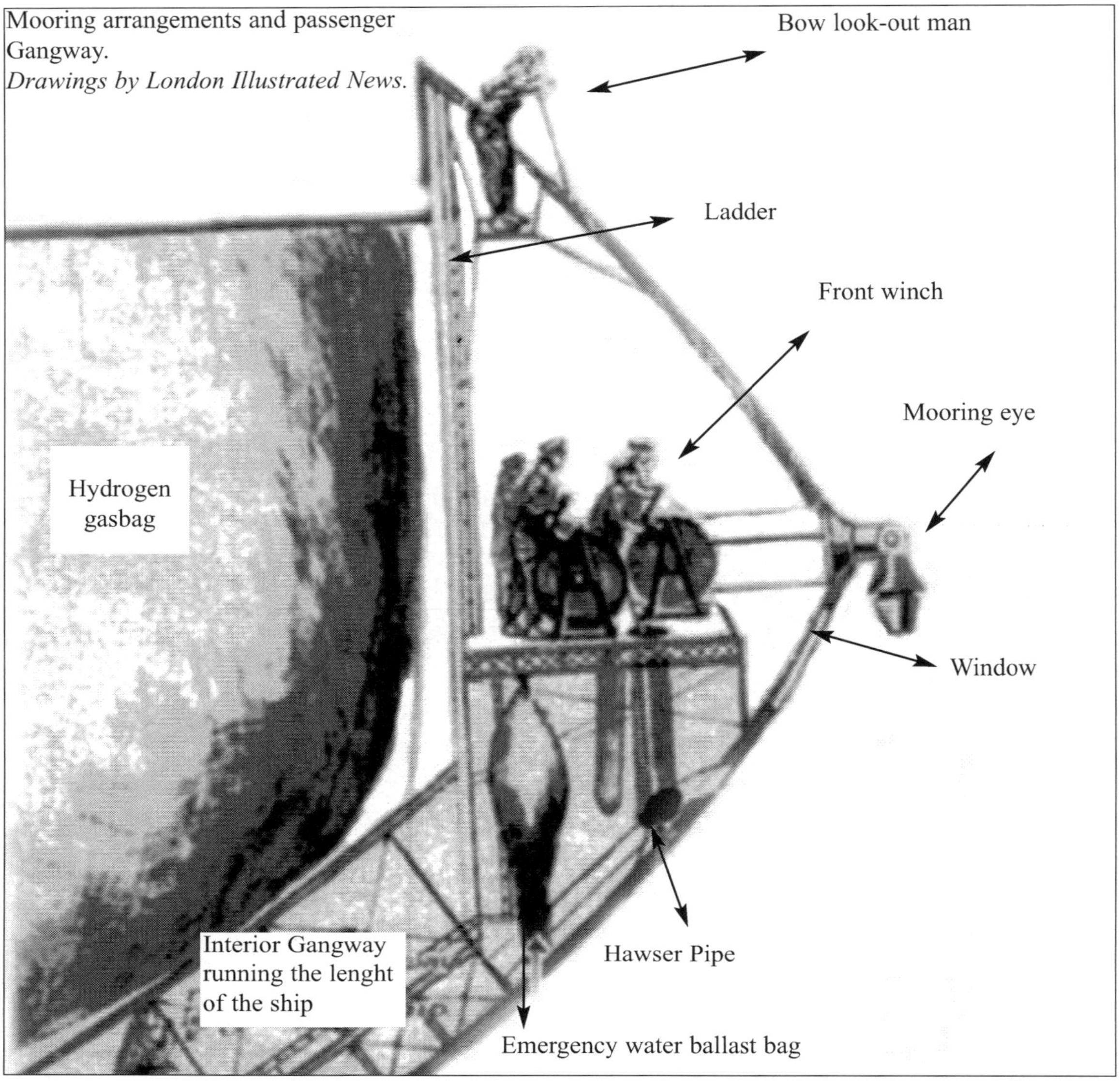

Mooring arrangements and passenger Gangway.
Drawings by London Illustrated News.

Some passengers boarding the *R-101* for a test flight. An observer supervises the manoeuvers from the nose of the dirigible.
Airship Heritage Trust Collection.

Société historique et culturelle du Marigot Collection.

Above London.

We get the impression that the *R-101* floats above the hangar of the *R-100*. It is only an optical illusion as the hangar is located about 800 metres from the tower.

Above: the original version of the *R-101*.
Below: the *R-101c*. The new extension is situated between the observation windows and the engines at the back of the airship.*Collection Flight*.

This photo is probably the one that best shows the part of the fabric that has not been replaced.
Collection Flight.

The two airships, the *R-100* in the foreground and *R-101* in the background, were usually kept one beside the other, but they rarely took off together. In this photo, taken at the time of the *R-100*'s last outing, we can hardly see beneath it all the people required to move it around. The *R-101*, at the beginning of October, 1930, is waiting for the signal to leave on its fatal voyage.

Lord Thomson of Cardington, Secretary of State for Air, Vincent Richmond, Assistant Director of Airship Development and Sefton Brancker, Director of Civil Aviation.
Airship Heritage Trust Collection.

One of the last photos of the crew, with Sefton Brancker, before boarding *R-101*. Almost all would die, some twelve hours later.
Airship Heritage Trust Collection.

The *R-101* in flight.
Photo Alfred G. Buckham.

One of the very last photographs of *R-101* before its fatal flight. Photograph at bottom: Lord Thomson, Secretary of State for Air, is about to take the elevator carrying him inside the airship.
L'Illustration nº 4571.

Above London, in 1929.

FATAL FLIGHT

POLITICAL PRESSURE

The *R-101*c was not ready for such a long trip. The exterior coating left a lot to be desired and the valves of the gasbags released hydrogen automatically whenever the airship pitched at an angle of more than three degrees. However, the Secretary of State for Air, Lord Thomson of Cardington, wanted to reach his goal of flying to Karachi within the schedule proposed in order to get back in time to attend the Imperial Conference which was to take place in London in mid-October. On July 17, he wrote: "I must insist on the program for the Indian flight being adhered to as I have made my plans accordingly." All of the authors do not agree on Thomson's responsibility; Peter Masefield is his most ardent defender.

After the installation of the new 14 metre middle section and a gasbag of a 510,300 cubic foot storage capacity, at the cost of £100,000, a test flight of at least 48 hours should have been undertaken, to confirm that the airship was capable of making such a long trip to India under very variable weather conditions. The test flight, on October 1st and 2nd, lasted barely 16 hours and 51 minutes. The *R-101*c flew over London. It weighed 117.9 tons with a payload 49.3 tons for a total of 167.2 tons. The *R-101*c was not subjected to any comprehensive testing nor to any test flight that would have simulated the real load for the trip to Britain's colony in India.

It was up to Colmore to make the final technical decision. His superior, John Higgins, who had just been replaced by Dowding, spontaneously acknowledged that he was not competent to make such a decision. A great deal of political pressure rested on Colmore's shoulders. If he delayed the trip, the Minister would probably shut down the enterprise. The social and political pressures, during a full blown economic crisis, did not make things any easier, especially at a time when the airplane was undergoing rapid and reassuring developments.

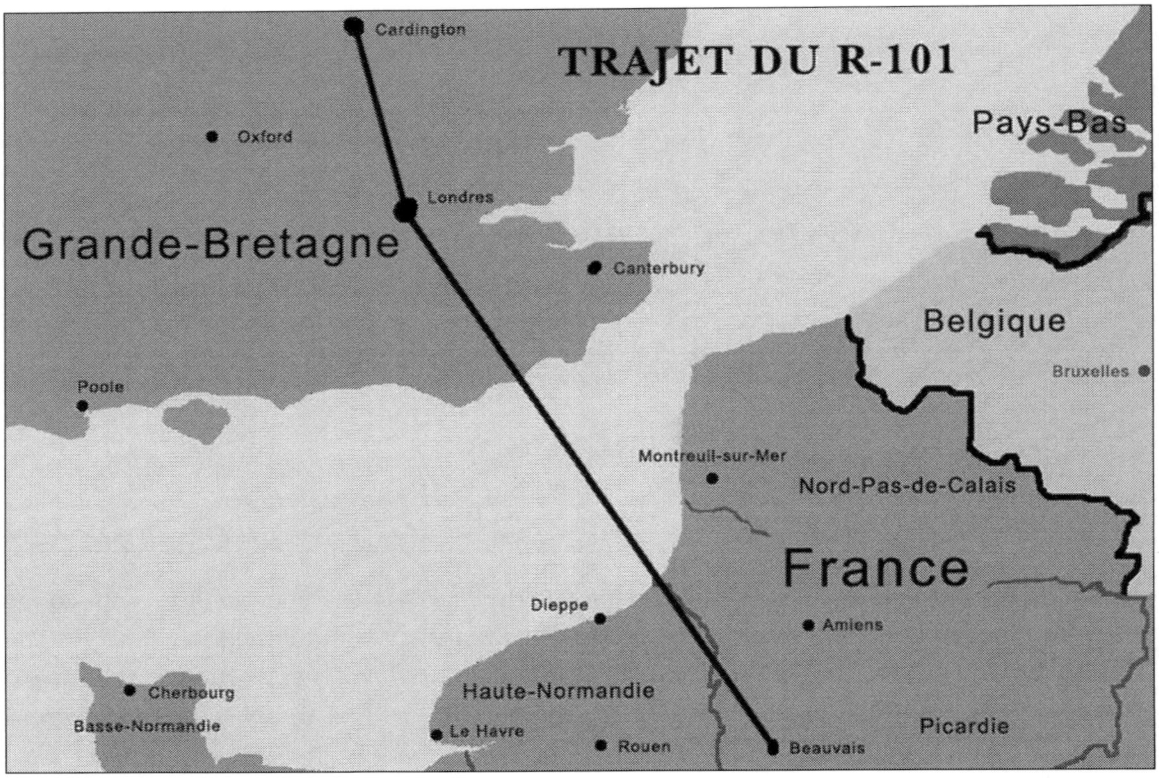

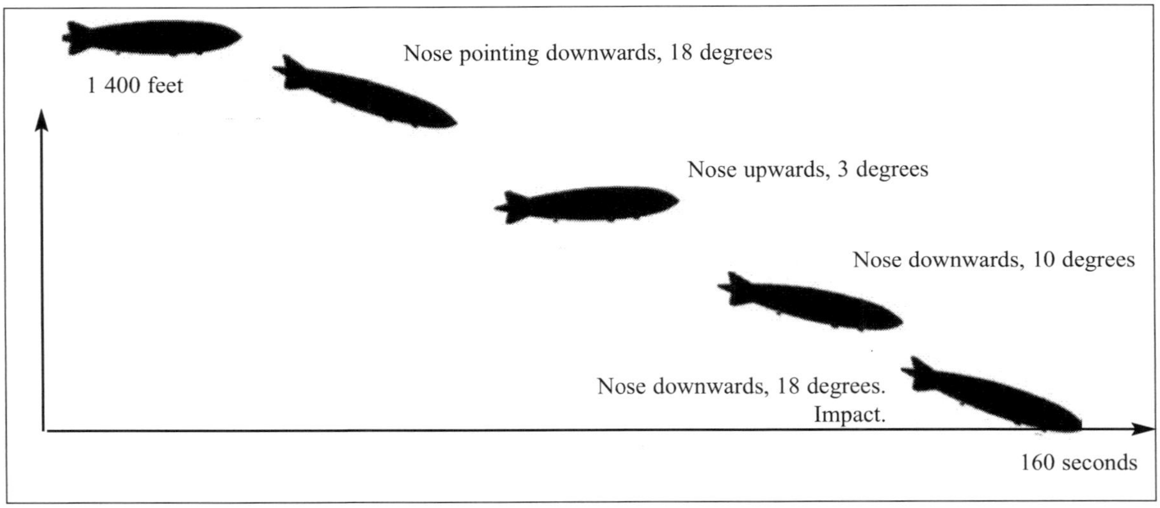

Source : *Airship Heritage Trust.*
http://www.airshipsonline.com

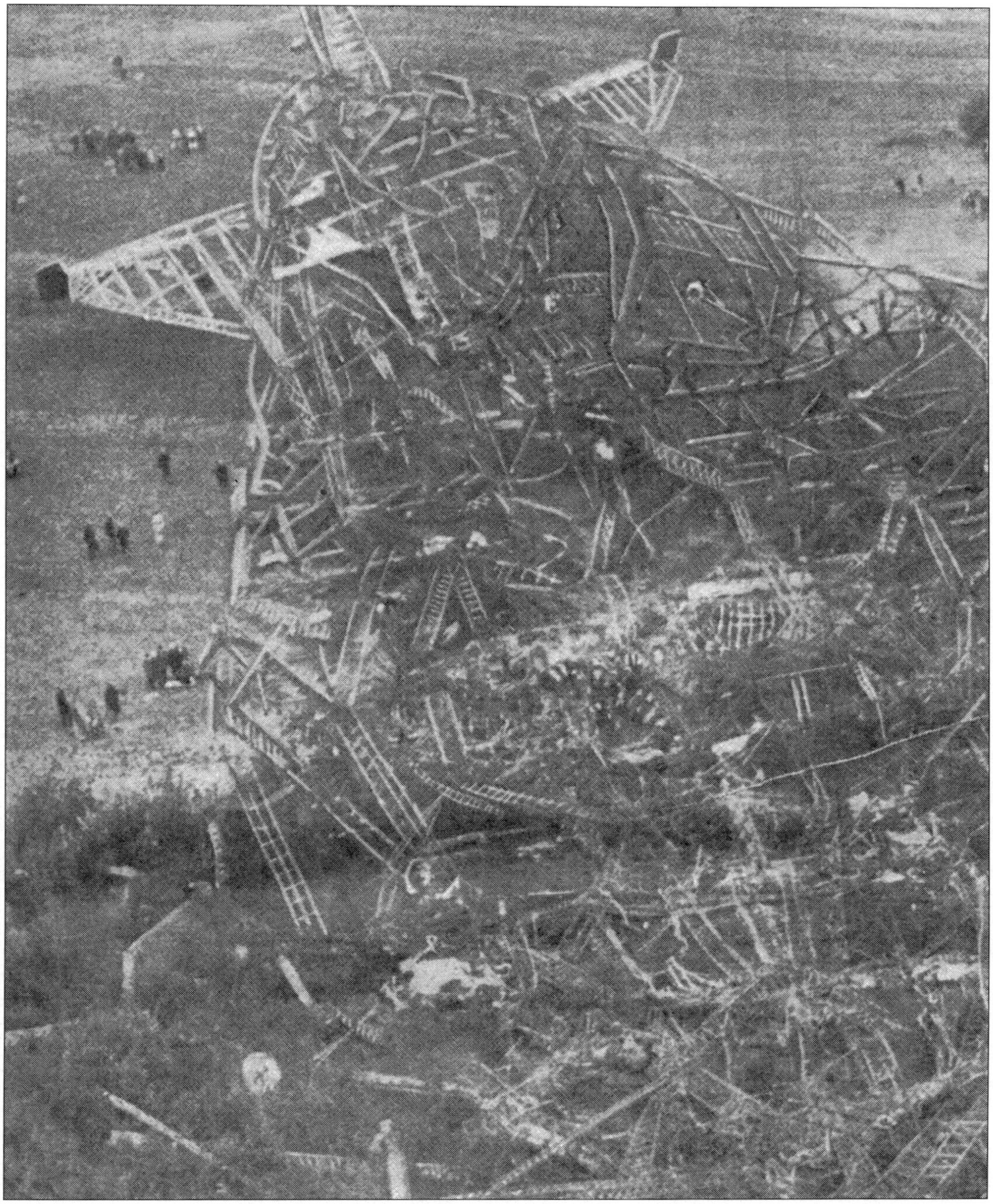

Overall view of the catastrop
Photo Daily Mirror.

The skeleton of *R-101* lying near the Fecq Woods.
Central Press Collection.

On the left, a drawing published in *Domenica del Corriere* of October 19, 1930. On the right, a drawing by S. Drigin, published in *Mid-Week Pictorial* of October 25, 1930.

Scene depicting the dramatic atrocity of the victims burned beyond recognition.
Photo Éditions des Portiques.

Corpses are covered with white sheets and aligned along a hedge.
Photo Daily Express.

Engine pod.
Photo Lelong. Amiens. Société historique et culturelle du Marigot Collection.

At the end of the tail of the ship, still flying, almost intact, the *R-101* flag. Notice in the sky a small plane, probably on a photograhic mission. The fabric has burned away but the framework seems to be in good condition. *L'Illustration nᵒ 4571.*

Front part of the dirigible with the nose pointing towards Paris.
L'Illustration n° 4571.

French soldiers on duty in front of one of the engine pods.
Times Wide World Photos et *Daily Mirror.*

Times Wide World Photos.

The flag of the R-*101,* held by French firefighters, remained almost intact.
Times Wide World Photos.

All that is left of the cockpit: the navigation wheels.
Times Wide World Photos.

Remains of one of the engines and of a valve that ensured the hydrogen pressure in the gasbags.
Times Wide World Photos.

French gendarmes searching for clues in the rubble.
Daily Mirror Collection.

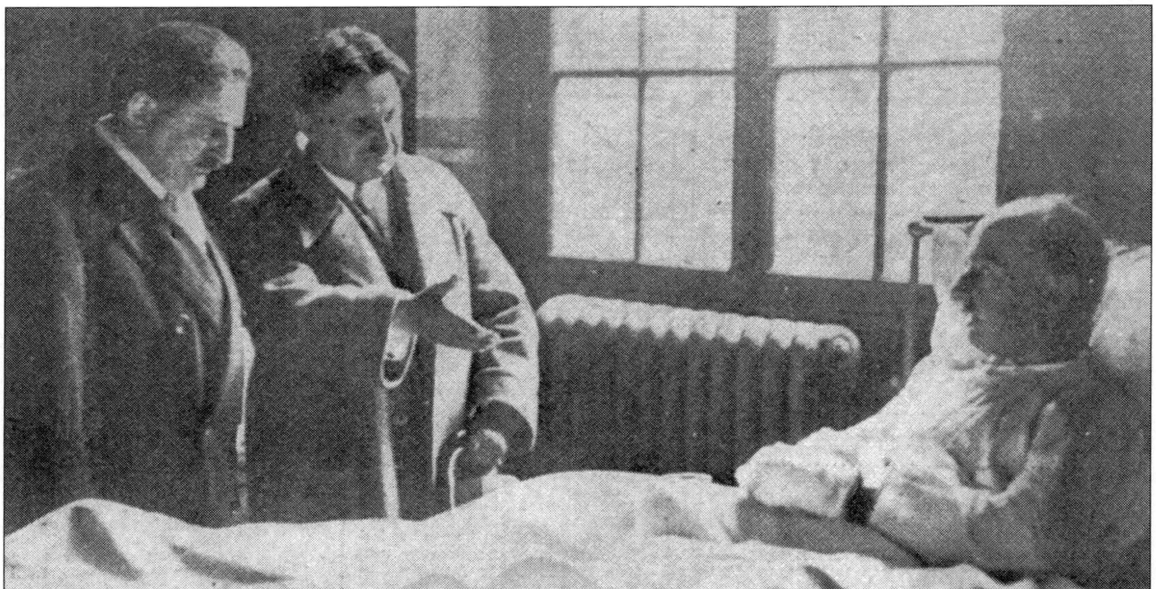

The visit of the French Air Minister, Laurent Eynac, talking with one of the survivors, A. J. Cook.
Daily Mirror Collection.

Minister Thomson had written that he was able to face up to the storm and declared that there was only one chance in a million that a person would be less safe in the *R-101* than in his own home. Colmore therefore confirmed that the airship was ready. His sole request, when he learned that Minister Thomson insisted on leaving as of October 3, the day after the test flight, was that the crewmembers be given a day of rest. The *R-101*c then left on October 4 for Karachi.

Few employees had agreed to such a hasty expedition, including Captain Irwin. Unlike the case of the *R-100*, whose workers had worked only a few years at Howden for the Airship Guarantee, the employees of the Royal Airships Works had been involved in the design and the construction of the *R-101* for many years. All the people in the village were therefore aware the airship was about to leave and many reunions were held here and there to highlight the departure of the expedition.

THE TRAGIC FLIGHT

During the takeoff, on a cold Saturday evening, October 4, the airship was so heavy that it had difficulty trying to gain altitude. Minister Lord Thomson of Cardington had insisted that a 15 by 9 metre Persian rug, from Kurdistan, be placed in the airship; the rug was to have been unrolled to receive the dignitaries at Ismailia, on the west bank of the Suez Canal. A case of champagne was added for that reception. Lord Thomson of Cardington himself had an impressive load of luggage, all while others pondered how to reduce the weight of the airship. Irregardless, the Secretary of State for Air took the elevator and forever left the soil of Cardington, the small town whose name he had adopted.

As one of the main engines was defective, numerous tests and three hours of repairs were required. At 6:36 p.m. the *R-101*c was detached from the mooring mast and lazily headed, not without difficulty, towards London. When it made a little nosedive, to elevate it, four tons of ballast were released, soaking many spectators. Only five tons of ballast remained for the whole trip. 25 tons of fuel were also stored on board. The silhouetted passengers and crewmembers, seen from the ground, waved to the spectators using flashlights.

At the outset, the airship was flying very slowly reaching hardly 32 kilometres per hour. Thirteen minutes later, the *R-101* was travelling at 65 km/h towards Ismailia, in Egypt, in spite of a poor weather forecast. While flying over London, the crew received a foul weather alert. The ship was going to be exposed to conditions the likes of which it had never encountered during any of the test flights. It was really audacious to consider making such a long distance trip under those weather conditions. The political ambitions and pride of Minister Thomson, who dreamed of being the next Viceroy of the British Empire in India, where he was born, proved to be very costly. The British people could very well have done without, some18 years later, a second Titanic. Not long before the tragedy, Lord Thomson of Cardington wrote to Princess Bibesco: “The *R-100* had given me pleasure, the *R-101* will give me, I hope, joy. To ride the storm has always been my ambition. Maybe we will achieve that on the way to India, but without unjustified risks for other human lives, at least I hope.” Lord Thomson of

Cardington had also told the *R-101*'s commander: "You must not let my natural impatience and my desire to depart influence your judgement." The *R-101*c flew over London around 8:15 p.m. and crossed the coast in the vicinity of Hastings at 21:36 just as the rain intensified. At 11:36 p.m., it passed above Point Saint- Quentin, on the coast of France near the north end of the Bay of Somme, at 296 kilometres from Cardington. At 2 a.m., the ship was moving at only 53 km/h over Beauvais, heading towards Paris, Tours, Toulouse and Narbonne.

The last radio message, received at Le Bourget, read as follows: "At present, the passengers, after eating an excellent meal and smoking cigars, having observed the French coast, are about to go to bed."

The airship, which barely got above 450 metres in altitude during its flight, flew with difficulty over the Beauvais region, passing only a few metres away from the magnificent cathedral. It was raining in torrents when the *R-101*c found itself in the midst of violent gusts of wind, almost a hurricane, at less than 100 metres from the ground. At 2:09 a.m. the airship crashed some five kilometres to the south-east of the town, after a seven and a half hour long flight. The rain, absorbed by the fabric, had considerably increased the weight of the apparatus. The impact with the ground, at a place called Les Coutumes, on the edge of Bois de Fecq, had not been very violent, yet it was enough to create a devastating fire. The airship went into a nosedive twice before hitting, pitched at a 20 degree angle, a small woods and a beet field where it immediately exploded. This cataclysm might well have been caused by the mixture of hydrogen with air, at a moment when one of theengines was still functioning. A few small explosions followed. The official report of the French police specifies: "[…] all of a sudden, a gigantic flash lit up the sky, at the same time a very loud explosion erupted, immediately followed by such a rush of air that some thought, that, finally, a powder magazine or an ammunition depot, or the Beauvais gasworks, had exploded." A few seconds after the explosion of the 5.5 million cubic metres of hydrogen, the *R-101*c was no more than a skeleton.

Albert-Eugène Rabouille. Les Éditions des Portiques.

Albert-Eugène Rabouille, a 57 year old button manufacturer, resident of Bongenoult, in the Commune of Allone, claimed to have witnessed it all. Although many persons from Beauvais had heard the roaring of the airships' engines and then the noise from the explosion, Rabouille was one of the only three eyewitnesses from outside the *R-101*c. The very same Rabouille had been poaching. He related to Jean Ajalbert: "The airship *R-101* was in distress - it turned out its lights twice - at the moment when it turned them back on, it crashed some 50 or 100 metres from me - I was under a tree - I fell on my behind in the mud - I heard three explosions - I saw big red flames. I also heard screams [...]" However, by the time the firefighters and the local police arrived at the scene, the witness had already taken a powder, returning only later to the crash site.

Princess Bibesco put it this way: "In the Allone Woods, Rabouille, the poacher, sets his trap. He lies in wait. But what he sees coming is neither the nocturnal hare, nor the wandering partridge, it is the fortunes of England being blown by the wind. The snare that entraps them, where they fall, it is that small hill standing against the sky since the beginning of time. That is where they crash: that is where, in one foul swoop, they find their funeral pyre."

The other witnesses were George Darling and Marcel Debaupuis. They watched as the *R-101* crashed and then rushed to provide assistance. They even entered the tail section to try to save passengers trapped inside. They received, for their gesture, a letter of commendation from the Government of France.

Louis Petit, owner of a general store, had also been witness to the *R-101*'s last moments. He reported that the airship was flying very low and that its lights were continuously turning on and off. Then, with all lights on, the airship went into a nosedive followed by a violent explosion.

THE SURVIVORS

As the crash of the *R-101* happened during the night, while almost all the passengers slept (only 12 people were on duty), there were only six survivors out of 54 in this tragedy south of Beauvais. The Secretary of Air, Lord Thomson, who had directed the construction of the *R-101*, was amongst the deceased. Two people who managed to make it out of the airship later died of their wounds. Sam Church survived for three days and even provided precious information to the investigators, and W.G. Radcliffe. The latter was declared dead by doctors at the hospital, but when he awoke, surprising everybody, the doctors realized their error; he had merely slipped into a coma. Radcliffe died however, a few hours later.

As for Church, he specified that he was about to release some ballast in order to elevate the dirigible when the accident occurred. He declared having heard a tremendous cracking noise above his head, as if a metal beam had split.

Mechanics Arthur V. Bell and John H. Binks, who were working in an engine-housing at the back of the airship, had survived due to the bursting of a water tank above their heads protecting them from being burned. Binks, a few minutes late for duty because he over slept, survived. He had been in discussions with his substitute, Arthur Bell, when the accident occurred. He had just barely slowed down the engine after he received a message by telegraph, from the cockpit.

Engineer Harry Leech, a burly man, was in the smoking-room where the smell of cigars lingered; he had been protected by the asbestos-covered walls of that room. He tore through the fabric, threw himself out through the flames and found himself stuck in a tree. Bell, Binks, and Leech had shaken hands and sworn to each other that they would make it out alive, a few seconds before leaving the airship.

Among the six survivors, four worked in the engine pods. These were aerodynamically shaped like an egg and suspended by slings under the hull, towards the outer edge of the dirigible, the least damaged part. Savory was in Engine Cabin Number 3, Cook, in Engine

Cabin Number 4, and Bell and Binks, in Engine Cabin Number 5. Leech was alone in the smoking-room. Disley was sleeping in his cabin, on the first deck. Journalist John Pitt told this anecdote concerning Arthur V. Bell: "Very nervous, Bell tried to calm himself by lighting a cigarette from the flaming carcass only to realize later that he had just lit a cigarette from what was turning out to be the most expensive lighter in the world." Others, in vain, tried to pass through what had quickly turned into a real net of steel.

RESCUERS

Alerted by the explosions, villagers rushed to the scene either on horseback, by bicycle or by car, guided by the bright flames dancing against an exceptionally dark sky. No one could really get close to the airship because the inferno was so intense. Even those with a burning desire to intervene to save some of the passengers or crewmembers were dissuaded. Some of the survivors, of course, tried, to help their comrades, but also just in vain. All they were able to do was to comfort them verbally for a few minutes. The wailing was unbearable to hear.

The news of the tragedy started to spread in the airports, in the pressrooms. Le Bourget transmitted the following information: "G-FAAW is on fire". British Prime Minister James Ramsay Macdonald told of Lord Thomson of Cardington: "During our last conversation, he had promised to send me two messages a day: one in the morning to tell how he had spent the night and the other in the evening, to tell me about his day. On Sunday, early in the morning, while I was still in bed reading by the light of a lamp, the phone rang. Thinking the call was from him, I answered joyfully. But it was not his voice, though the news was about him: I was informed that two hours earlier, the *R-101* had crashed, near Beauvais, and that my friend was dead."

Cameraman Paul Wyand, of Movietone, the company that produced the news coverage for British movie houses, received a phone call at 5:30 in the morning. His boss informed him that he had to take a charter plane, to Croydon, as soon as possible. He grabbed his camera and drove off in his Amilcar, his small French sports car. Wyand had been ordered to bring back the "rushes" in time to allow the company to show them in movie theatres the very same evening. Arriving over Beauvais he began filming as soon as he saw the metal carcass on the ground. He arrived at the scene of the accident at 8:30. There he recorded scenes that made him sick. He had a lot of difficulty overcoming his emotions in the face of policemen and firefighters darkened with soot and villagers in tears. He left by 10 o'clock. The scenes he had filmed were screened in London theatres by 6:30 p.m. The remains of the airship looked like the skinned carcass of a gigantic whale with its crushed head pointing towards Paris. The centre section, the one that had been added on, had almost disappeared while the metallic envelope at the back was almost intact. The flag which included part of the British flag, attached to the end of the tail, was found completely intact, probably saved by the rain. It is presently stored in one of Cardington's churches.

Three of the survivors back at the crash site: A.V. Bell, H.J. Leech and J H. Binks.
Daily Mirror Collection.

Citizens of the town of Beauvais arrive at the hospital to get the latest news.
Daily Mirror Collection.

The French authorities deal with the transport of the victims. Note that certain coffins are particularly small.

Transporting the coffins on a cart.
L'Illustration n° 4571.

Five coffins deposited on the ground adjacent to the small Town Hall of Allonne.

The officers and the soldiers did their job meticulously, under quite difficult conditions, especially because of the crackling sounds of the huddled up corpses that were being carbonized and reduced to the size of children. The heat was so intense that rings on the fingers of crewmembers had melted. Bones as well as pieces of flesh were brought out, but at times only to fall apart or crumble to the ground. The investigation report of the French police was even more explicit: "the flesh on those bodies was even cooked, carbonized, and yet some of the rescuers had to extirpate flesh and even entrails stuck to thousands of intertwined steel wires which covered most of the bodies."

The identification of the bodies was difficult; they all looked alike. In one case, it was believed that an extra body had been omitted from the official list, but it was later that it was in fact one body, separated in two. At first bodies were placed along a long hedgerow at the edge of the forest; later, a white sheet was placed over the heads and still later, over the whole body. It was the people of Allone who provided those sheets and who placed flowers on the victims' caskets. Jean Mortane wrote: "Once more, with pain, hearts beat together. People understood each other. Might they never forget." The catastrophe generated scenes of solidarity as spontaneous as they were improvised.

A journalist from the *Journal*, Émile Condroyer, described the scene:

"Shrouds bunched up by extreme contortions, feet striking out, mangled and black, like dead branches. Extricated from the mud with great effort, the funeral cart arrived. Whenever policemen with rubber-gloved hands took up, one by one, the miserable remains, they could not help notice that the remains no longer resembled human beings. Sometimes the bones were discolored, like old ivory. Every time one of the charred corpses was placed in a coffin, it cracked like a twig."

The most charred bodies, and particularly 22 unidentifiable ones, all came from the centre part of the airship where the fire started. Personal items were placed in white caskets that were arriving progressively as the local capacity, of the Maison Jouvin, of Beauvais, was insufficient. The caskets were placed in a small school, but quickly that space was filled up and a dozen corpses had to be put under guard.Some corpses were placed in a temporary morgue in the small Allone town hall.

During the evening of October 5 and on the night of October 6, some sisters from a religious community visited the dead and prayed for salvation of the victims' souls. The caskets were then transported to the Beauvais city hall.

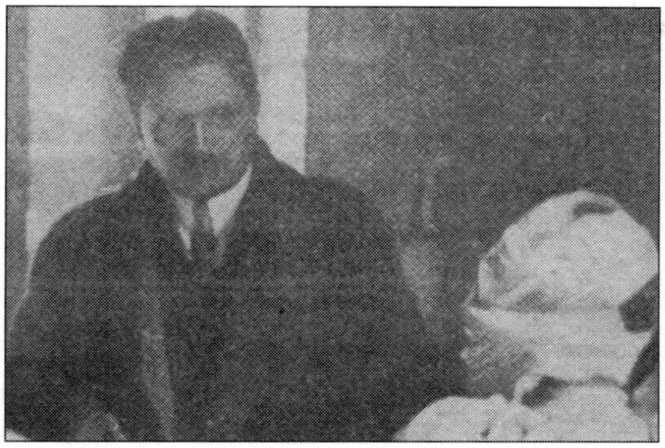
French Air Minister André Laurent-Eynac.
Photo Daily Mirror.

French Air Minister, André Laurent-Eynac, arrived at the crash site and then went to comfort the wounded at the Beauvais hospital. Amongst them was W.C. Radcliffe, whose severe burns brought on his death hours later. Jean Ajalbert arrived during the morning. His testimony was moving: "When one looks away, when one is removed from that horrendous and macabre carcass, all one can see in the distance, across the curved plain, are endless lines of ants: on the roads, endless lines of cars, on the paths, the movement of pedestrians, and, in the sky, planes coming to pay their respects to the extraordinary remains. In the same manner that I saw camel skeletons, along the paths, in some Moroccan village … and those planes, diving, turning, amicably, for a moment, make me think of raptors about to fall upon some exhausted animal …"

A correspondent from the Daily Express, H.J. Greenwall, met a colleague from the periodical Le Matin, the photographer, M. Mathieu. "Ah, my poor buddy! This is the most dreadful moment of my life.", he said and then burst out sobbing. That was, according to Greenwall, the most difficult reporting assignment of his career.

People were pushing and shoving each other; some looking for a souvenir, others meticulously scanning the site for small items such as spoons or items as big as propellers, or simply pieces of aluminum from the hull or even metal wires which would later serve to make fences.

THE FUNERALS

Such a slaughter inevitably brings its share of condolences. Expressions of sympathy came from many parts of the world, from American President Hoover to the Italian Air Minister, Italo Balbo.

The President of the Council of France, André Tardieu, declared the 7th of October a day of national mourning and flags flew at half-mast. Many businessmen from Beauvais closed their shops in order to attend the funerals.

The official funeral ceremony took place at Beauvais City Hall. For the occasion, an enormous funeral hanging with the British flag in the center had been attached to the facades of the carriage entrance. In addition to the local personalities, the President of Council, André Tardieu, Air Minister, André Laurent-Eynac, and the British Ambassador, Lord Tyrell, attended the ceremony. The same morning, a violent hailstorm fell upon Beauvais.

At the start of the funeral ceremony, a 101 gun salute, in keeping with tradition and not because of the name of the airship, was paid in tribute. The British National Anthem, God Save the King, was played and then the caskets were loaded, draped with the British flag, each in chariot pulled by four horses. Thirty French airplanes and a British squadron criss-crossed the sky and flew over the cortege. For a few years thereafter British planes continued to fly over Allone, on October 5th, while people from the community placed flowers on the crash site.

The funeral procession to the station was impressive. Overwhelmingly, a crowd estimated at 30,000 people, followed the remains from the city hall to the station. Many French and British flags flew at half mast, covered in black crepe. On Jeanne-Hachette Square, between the town hall and the monument dedicated to Jeanne Hachette, some soldiers carried large floral wreaths. Cavalry troops, mounted on small horses, stood at attention.

The most touching scene was the arrival of three of the survivors: Arthur Bell, John Binks and Harry Leech. Soldiers escorted the funeral convoy, rifles pointed at the ground. Following the fly past, all that could be heard were the wheels of the 24 chariots and the hoofs of the horses on the pavement.

After having arrived at the beautiful little neo-medieval train station, of Petite Vitesse, at Beauvais, the 51st Infantry Regiment played the Last Post. The caskets were then loaded onto seven railroad coaches. At Boulogne-sur-Mer, the caskets were hoisted onto the British destroyer Tempest, a name reminiscent of the circumstances surrounding the crash of the airship. Early in the evening of October 8th, the destroyer reached the Port of Dover. Once more the caskets were transferred to a train, and then taken to the Westminster morgue. On the same day, the death of Samuel Church was reported, bringing the number of victims to 48. His body was taken to London the next morning.

On Friday, October 10, the caskets were exposed in the Hall of Westminster Palace upon a giant catafalque. Numerous floral wreaths had been placed in front of and surrounding the caskets, in two straight lines. Tens of thousands of people marched-passed from eight in the morning until midnight.

At noon, religious funerals were performed at the St Paul's Cathedral. Many dignitaries attended the ceremony, including the British Prime Minister, James Ramsay Macdonald, the Prince of Wales, the Duke of York, the Lord Mayor of London, the Italian Air Minister, Italo Balbo, and his French counterpart, André Laurent-Eynac, as well as the survivors, Bell, Binks and Leech. For the occasion, the flag of the airship was placed on the altar.

The music was performed by a group from the Royal Air Force, under the direction of J.H. Amers. They played:

Marche Solennelle Gounod
Largo Haendel
Funeral March Mendelssohn
Airships R-100 and R-101
Ave Verum Mozart
Solemn Melody H. Walford Davies

Follows extracts of the hymn chosen for the occasion. Note the references to the storm (tempest) and flight (fly):

Jesus, Lover of my soul, Let me to thy bosom fly,
While the nearer waters roll,
While the tempest still is high:
Hide me, O my Savior, hide,
Till the storm of life be past!
Safe into the haven guide,
Oh, receive my soul at last!

After the blessing, they played the *Dead March* and the *Last Post*.

A mass was celebrated for the four deceased Catholics at the Westminster Cathedral, by Father Harry Rope, brother of Squadron Leader Frederick M. Rope, who died in the catastrophe.

The funerals took place on the following morning, October 11th. Present, in addition to those personalities of the day before at St Paul's Cathedral, was Dr Hugo Eckener, airships specialist, who had made the first world tour on a dirigible, the Graf Zeppelin. The bodies were transported by train to Bedford. All the shops there were closed for the occasion. Then they were taken to the St. Mary of Cardington cemetery where they were all interred in a mass grave at the end of the day. The crewmembers of the *R-100* attended that final ceremony which ended with the poignant notes of the *Last Post*. A commemorative plaque had been erected on the site. It reads: "Here lie the bodies of 48 officers and men who perished in H.M. Airship *R-101* at Beauvais, France Octr 5th 1930". The flag of the *R-101* has been hanging in the church of St. Mary since September 28, 1931.

The widow of Frederick Michael Rope, Lucy Jolly, had a Catholic church constructed on the family farm, at Kesgrave, near Ipswich. One can find inside a replica of the *R-101*. There were 54 people on board the *R-101* when it left, but only six returned. The only consolation for some of the families, a few of the passengers had life insurance. Back then, despite the risks, it cost only 1% of the principal sum for the insurance. Lord Thomson of Cardington had thus insured himself for £500 at a cost of £5; he had left a note on his desk indicating that he would be bequeathing all his property to his brother in the case of an accident with the *R-101*. The airship itself was not insured.

HONOURS

On November 17, the British ambassador in Paris delivered a letter to Foreign Minister Aristide Briand, testifying to his government's gratitude for the assistance provided during the tragic events of October 5.

Three of the six survivors of the tragedy of *R-101* attend the funeral ceremonies held according to protocol in Beauvais: Joe H. Binks, wearing a pale coat, Harry J. Leech, in the center, and Arthur V. Bell. In front of them, the coffin of one of the victims. All prepare for the departure to the station at Boulogne.

In Beauvais, Spahi Cavalry render a military salute with their glistening sabres.
Daily Mirror.

Central News. Société historique et culturelle du Marigot Collection.

The convoy in front of the Beauvais Town Hall. On horseback, in a pale uniform, the Spahis of Senlis render the honours. Photograph below: arrival at the Beauvais Station.
Société historique et culturelle du Marigot Collection.

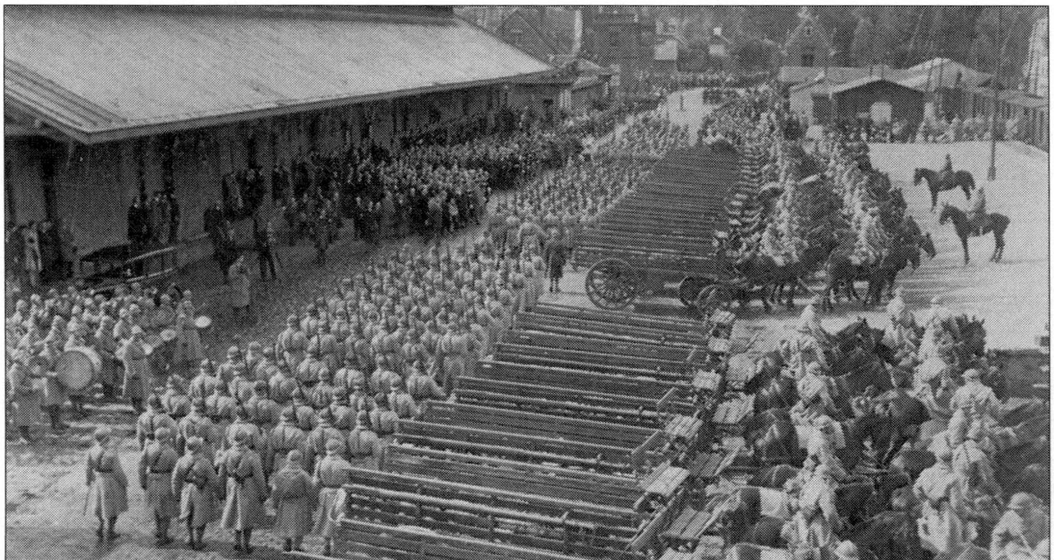

Representatives of the French and British governments, at the Beauvais Station, October 7, 1930. In the center of the first row, the survivors of the tragedy: Joe H. Binks, Arthur V. Bell and Harry J Leech (head bandaged). In the second row, on the right: the British General Salmon, André Tardieu and Laurent-Eynac.

Transporting the coffins on board the *Tempest*. In the background, some buildings of the town of Boulogne.
L'Illustration nº 4572.

October 7. The caskets aboard the Tempest at Boulogne.
The Topical Press Collection.

The flag of *R-101* exposed upon the altar of St Paul's Cathedral.
Times Collection.

Coffins in Westminster Hall.
Times Collection.

Quiet homage paid by thousands of people filing past in front of the coffins. On the left, a soldier of the Royal Air Force mounting the guard, weapon pointed towards the ground.
Daily Mirror Collection.

Members of the Royal Air Force leading the funeral procession after the religious service celebrated at St Paul's Cathedral.
Daily Mirror Collection.

The funeral procession leaves Strand Street and takes Aldwych Street, in London.
Daily Mirror Collection.

The Italian Aviation Minister, Italo Balbo.
Daily Mirror Collection.

Prime Minister Ramsay Macdonald, accompanied by his son.
Daily Mirror Collection.

On the right, the French Air Minister, Laurent-Eynac.
Daily Mirror Collection.

The designer of the Graf Zeppelin, Dr Hugo Eckener. First on the left, Dennistoun Burney, designer of the *R-100*.
Daily Mirror Collection

The procession leaves Whitehall. Postcard.
Société historique et culturelle du Marigot Collection.

The procession of the 48 coffins on artillery wagons passes via Charing Cross Station.
Photo Flight.

The silent crowd watches as the procession passes, heading from Bedford to Cardington.
Daily Mirror Collection.

Gun salute.
Daily Mirror.

The coffins are placed in the common grave of the cemetery of Cardington, near the airships base. The last rites, shared by ministers of various denominations, will follow.
Photo Union.

Memorial over the single mass grave in Cardington cemetery.
Société historique et culturelle du Marigot Collection.

On December 3, at the Beauvais Town Hall, the rescuers from Beauvais, Allone and Voisinlieu were rewarded by both the French and the British governments.

ORDER OF THE BRITISH EMPIRE

The Commander of the British Empire (CBE) medal has only been given to a very limited number of people: Louis Couhé, Secretary General of the Air Ministry and Georges LeBeau, Prefect of Oise. That distinction was created in 1917 by George V. It is described as the "Most Excellent Order of the British Empire" and includes reference to "For God and the Empire". That honour confers upon its recipient the status of nobility, that is, a knighthood. There is no star for the Commander, and the insignia attached to the ribbon (6.35 centimetres wide) of the Commander is smaller than the one for the Knight of the British Empire (KBE).

This decoration is described in the following manner:

Obverse: a silver-gilt cross of four members with three points each, enamelled in translucent pearl grey surmounted by the Imperial crown. In the circular centre, in gold, Britannia is represented holding a trident, sitting next to a shield and bearing the national flag. She is surrounded by a crimson enamel ring upon which is inscribed the motto "For God and the Empire" in gold letters.
Reverse: a circle of rope cord marked with the royal number of the period, surmounted again by the Imperial crown and punch marked in the lower section. A large ring is passed through the smaller ring at the top of the crown to receive the purple 3.8 centimetre-wide ribbon.

THE ORDER OF THE COMPANIONS OF HONOUR

Decoration established 4 June 1917 by King George V at the same time as the Order of the British Empire. This medal is awarded for a contribution of national importance. It is an oval-shaped medallion, in gold. One finds upon it an oak tree. The Royal coat of arms hangs from one of the branches. On the right side there is a knight in armor mounted on a horse. The following is inscribed: "In action faithful and in honour clear". This honour was given to the Mayor of Allone, Félix Ballon, to the Mayor of Beauvais, Auguste Joly, to Squadron Leader Serin, Commander of the Gendarmerie of Oise and author of the official investigation report, and to Colonel Langlois, Commander of the Beauvais District.

MEMBER OF THE ORDER OF THE BRITISH EMPIRE (MBE)

The honourees receive a medal almost identical to the one for Commanders. The member wears, on the left side of the chest, a non enamelled five centimetre-wide silver insignia.

This honour was bestowed upon Captain Bertschy and Captain Royol, both from the Gendarmerie of Oise, on Lieutenant Huprelle, a Beauvais firefighter, on Commander De Witasse, on the Garrison Major at Beauvais, and upon André Chenard, Clerk at the Oise Prefecture.

Medal of the Order of the British Empire for Meritorious Service

The medal, created in 1922, was not intended for any of those members of any one the five levels of the Order of the British Empire, nor for anyone who might later be decorated. It is a thin circular silver medal of 3.6 centimetres in diameter.

Obverse: Britannia is shown seated, the sun shinning on her right and bearing the legend: "For God and the Empire" and inscribed below: "For meritorious service".

Reverse: the Royal Cypher surmounted by a crown with the words appearing on two lines: "Instituted by King George" at the bottom within a border of four heraldic lions. The bar fastener bears laurel leaves affixed to the medal by a single claw.

Ribbon: the civic ribbon is purple (3.75 centimetres in width).

This decoration was presented to the following persons:
André Deboudt, Police Adjudant of Beauvais
Jerôme Dutel, Beauvais policeman
Henri Huprelle, Beauvais firefighter
Marius Lejosne, of the 51st Infantry Regiment
Marcel Lesigne, Squadron Sergeant-major of Beauvais
Joseph Levoir, Republican Guard, of Beauvais
Jean Plomion, Mechanic of Allone
Gaston Sauvage, Sergeant of the Allone firefighters
Maurice Vaudeville, Beauvais policeman

The British Embassy, in France, also presented a letter rolled inside a thick paperboard cylinder along with a silver cup adorned with the following inscription: "*R-101* Allonne (Drawing of the Royal coat of arms of England), October 5th 1930". That cup had been produced by the company R & D, of Birmingham.

The British government presented to the Commune of Allone a crystal inkwell on a silver platter. The Mayor of Allone also received a gold watch. The largest piece was a silver vase offered to the City of Beauvais. Moreover, about a hundred persons from the Beauvais region received various gifts and a letter from the British ambassador in Paris, Sir William George Tyrrell. That letter read:
"I have been instructed by my government to express to you its deepest gratitude for your valorous actions during the catastrophe of the airship *R-101*. I am charged with the responsibility of expressing my government's profound recognition for the spontaneous efforts that you made during those painful times.
My compatriots will always remember the gestures of sympathy shown towards them by the inhabitants of
Allone and Beauvais during the profound mourning that has afflicted my country. Very truly yours."

The French Interior Minister, André Tardieu, also bestowed various honorary distinctions, more or less upon the same people.

Médaille d'Honneur vermeil for bravery and devotion presented to Lieutenant Huprelle, firefighter from Beauvais.

Médaille d'Honneur d'argent for bravery and devotion presented to Jérôme Dutel, policeman from Beauvais.

Médaille d'Honneur de bronze for bravery and devotion presented to Félix Ballon, André Chenard, André Deboudt, Henri Huprelle, Marius Lejosne, Marcel Lesigne, Joseph Levoir, Jean Plomion, Gaston Quenot, Gilbert Renaud and Maurice Vaudeville. The Minister also conferred honorary mentions upon Joseph Deloge, of Voisinlieu, Lucien Lancaster and Léon Legrand, firefighters from Beauvais, and Robert Hoëppe, of Voisinlieu. Furthermore, some ten letters of commendation were given to some other citizens.

A stele of about one metre in height, bearing simply the inscription "Dirigible *R 101*, October 5 1930", was erected at the edge of the Fecq woods. In October 1933, an imposing commemorative monument was inaugurated beside Route nationale 1, in the region of Warluis. On its base, one reads: "To the memory of the victims of the catastrophe of the *R-101* Oct 5 1930. This monument was erected by Great Britain and France." On the front of the stele, are inscribed only the names and titles of Lord Thomson of Cardington, Sefton Brancker, Colmore, Richmond, Scott and Irwin; the names of the other deceased persons appear on the three other sides. This monument was moved about one hundred metres, in the 1990s, in order to build a roundabout.

INVESTIGATION REPORTS AND FINAL ASSESSMENT

All told, the *R-101* had made over a period of 16 days, 12 flights for a total duration of 127 hours and 20 minutes. It had remained 43 days moored for 1,039 hours and 53 minutes. The costs for the *R-101* were much higher than than those for the *R-100*. The budget of the *R-101* came to £711,000 versus £500,000 for the *R-100*: a 30% difference.

A French investigative commission from Allone focused solely on writing a report on the incident. This can be consulted, in its entirety, in the Appendix.

A public British investigative commission lasted for 13 days in London. The investigators questioned 42 witnesses. The report suggested that the primary cause of the accident had been the explosion of one gasbag at the front of the airship. Those gasbags were no longer held by the original cabling system and could therefore be continuously rubbing against the wire meshes. Once the hydrogen was completely released, the *R-101* went into a nose dive and hit the ground with an enormous impact; at that precise moment the probability of a fire had been multiplied by ten. Moreover, a great amount of water had fallen upon the fabric of the dirigible considerably weighing down the *R-101*. Finally, one should not forget that it had become impossible to release ballast from the cockpit as it had already been largely jettisoned during the takeoff.

Other factors could have caused the accident. The two-year old fabric was replaced during the elongation of the *R-101*c, but the part on the nose had only been patched. That howeverwas one one of the parts that experienced the most turbulence and «stress». Another important point is that the accident occured during the changing of the guard. The two best officers, Irwin and «Grabby» Atherstone, had just been replaced by Maurice Steff, much less experienced, and by Hunt, renowned for his competence, but who was probably very tired after all the preparations for the departure. Ultimately, the order given by the control cabin to switch the engines to neutral, was fatal. The speed thus dropped to almost nothing and raising the nose became almost impossible.

The crash of the *R-101* undermined the confidence of passengers. Furthermore the mooring towers at Ismailia and at Karaichi were never used.

Photo taken in 1933.
Photo Keystone.

The memorial at its new site.
Photo Pierre Gilbert (Paris).

Details of the monument.
Photo Pierre Gilbert (Paris).

The cairn is particularly difficult to find. Few people know its exact location. Originally, the inscription was engraved on the stone.
Photos Pierre Gilbert (Paris).

A few souvenirs of the R-101

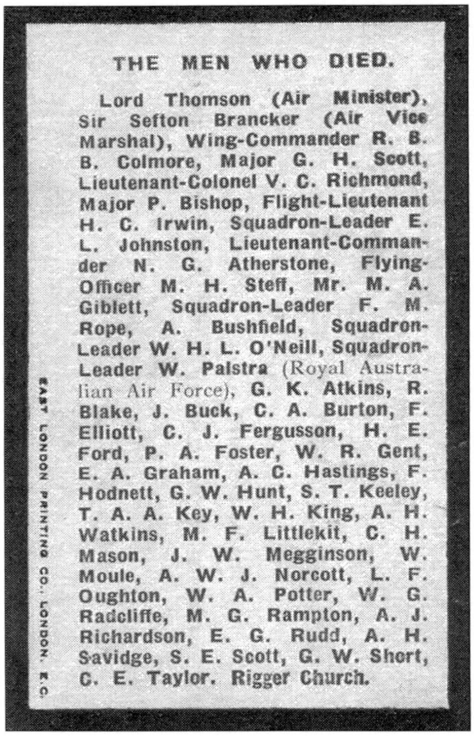

THE MEN WHO DIED.

Lord Thomson (Air Minister), Sir Sefton Brancker (Air Vice Marshal), Wing-Commander R. B. B. Colmore, Major G. H. Scott, Lieutenant-Colonel V. C. Richmond, Major P. Bishop, Flight-Lieutenant H. C. Irwin, Squadron-Leader E. L. Johnston, Lieutenant-Commander N. G. Atherstone, Flying-Officer M. H. Steff, Mr. M. A. Giblett, Squadron-Leader F. M. Rope, A. Bushfield, Squadron-Leader W. H. L. O'Neill, Squadron-Leader W. Palstra (Royal Australian Air Force), G. K. Atkins, R. Blake, J. Buck, C. A. Burton, F. Elliott, C. J. Fergusson, H. E. Ford, P. A. Foster, W. R. Gent, E. A. Graham, A. C. Hastings, F. Hodnett, G. W. Hunt, S. T. Keeley, T. A. A. Key, W. H. King, A. H. Watkins, M. F. Littlekit, C. H. Mason, J. W. Megginson, W. Moule, A. W. J. Norcott, L. F. Oughton, W. A. Potter, W. G. Radcliffe, M. G. Rampton, A. J. Richardson, E. G. Rudd, A. H. Savidge, S. E. Scott, G. W. Short, C. E. Taylor. Rigger Church.

EAST LONDON PRINTING CO., LONDON, E.C.

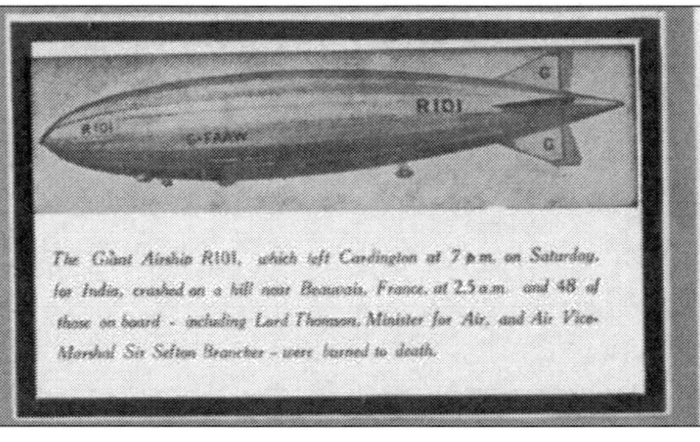

The Giant Airship R101, which left Cardington at 7 p.m. on Saturday, for India, crashed on a hill near Beauvais, France, at 2.5 a.m. and 48 of those on board - including Lord Thomson, Minister for Air, and Air Vice-Marshal Sir Sefton Brancker - were burned to death.

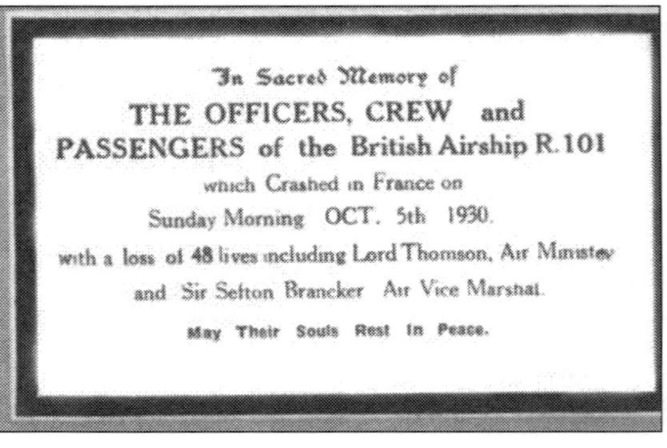

In Sacred Memory of
THE OFFICERS, CREW and
PASSENGERS of the British Airship R.101
which Crashed in France on
Sunday Morning OCT. 5th 1930.
with a loss of 48 lives including Lord Thomson, Air Minister
and Sir Sefton Brancker Air Vice Marshal.
May Their Souls Rest In Peace.

Société historique et culturelle du Marigot Collection.

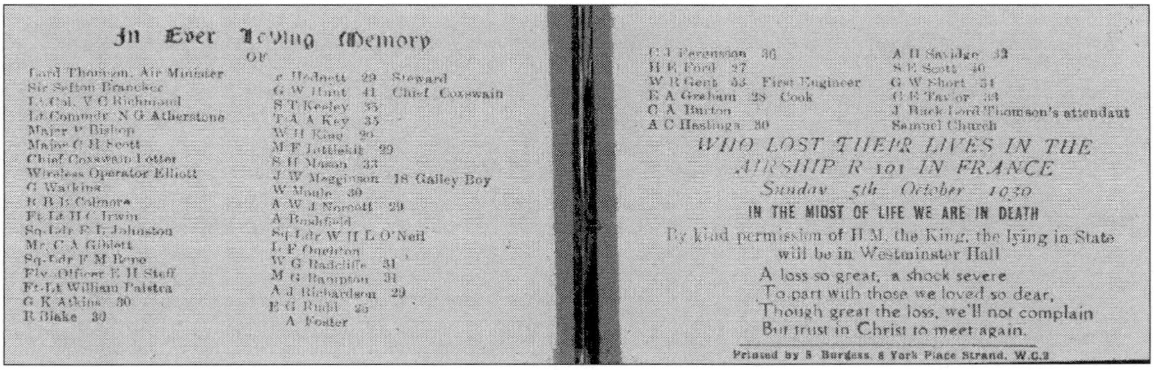

In Ever Loving Memory
OF

Lord Thomson, Air Minister
Sir Sefton Brancker
Lt-Col. V C Richmond
Lt-Commdr N G Atherstone
Major F Bishop
Major G H Scott
Chief Coxswain Potter
Wireless Operator Elliott
G Watkins
R B B Colmore
Ft-Lt H C Irwin
Sq-Ldr E L Johnston
Mr. C A Giblett
Sq-Ldr F M Rope
Fly.-Officer E H Steff
Ft-Lt William Palstra
G K Atkins 30
R Blake 30

A Hodnett 29 Steward
G W Hunt 41 Chief Coxswain
S T Keeley 35
T A A Key 35
W H King 20
M F Littlekit 29
S H Mason 33
J W Megginson 18 Galley Boy
W Moule 30
A W J Norcott 29
A Rushfield
Sq-Ldr W H L O'Neil
L F Oughton
W G Radcliffe 31
M G Rampton 31
A J Richardson 29
E G Rudd 25
A Foster

C J Fergusson 36
H E Ford 27
W R Gent 53 First Engineer
E A Graham 28 Cook
C A Burton
A C Hastings 30

A H Savidge 32
S E Scott 40
G W Short 31
C E Taylor 33
J Buck Lord Thomson's attendant
Samuel Church

WHO LOST THEIR LIVES IN THE AIRSHIP R 101 IN FRANCE

Sunday 5th October 1930

IN THE MIDST OF LIFE WE ARE IN DEATH

By kind permission of H.M. the King, the lying in State will be in Westminster Hall

A loss so great, a shock severe
To part with those we loved so dear,
Though great the loss, we'll not complain
But trust in Christ to meet again.

Printed by S Burgess 8 York Place Strand, W.C.2

Société historique et culturelle du Marigot Collection.

Commemorative medal fabricated by J.R. Gaunt, Birmingham.
Société historique et culturelle du Marigot Collection.

THE KING'S MESSAGE

I am horrified to hear of this nationa disaster which has befallen airship R 101, and the consequent loss of life, including that of Lord Thomson my Air Minister. The Queen and I sympathise deeply with the relatives and friends of those who have perished in the service of their country and also with the injured survivors.

GEORGE R.I.

Thy Will be Done,

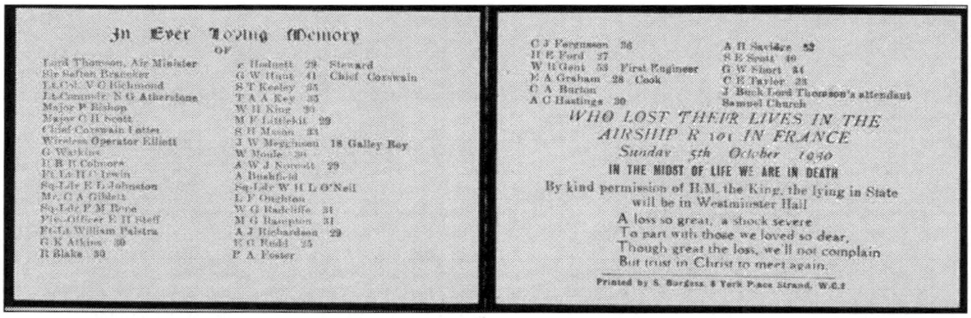
In Ever Loving Memory
OF

Lord Thomson, Air Minister
Sir Sefton Brancker
Lt.Col. V C Richmond
Lt.Commdr. N G Atherstone
Major F Bishop
Major G H Scott
Chief Coxswain Latter
Wireless Operator Elliott
G Watkins
R B B Colmore
Ft. Lt. H C Irwin
Sq-Ldr E L Johnston
Mr. C A Giblett
Sq-Ldr F M Rope
Flo.-Officer E H Steff
Ft.-Lt. William Palstra
G K Atkins 30
R Blake 30

[illegible] Hodnett 29 Steward
G W Hunt 41 Chief Coxswain
S T Keeley 35
T A A Key 35
W H King 30
M F Littlekit 29
S H Mason 33
J W Megginson 18 Galley Boy
W Moule 30
A W J Norcott 29
A Bushfield
Sq-Ldr W H L O'Neil
L F Oughton
W G Radcliffe 31
M G Rampton 31
A J Richardson 29
E G Rudd 25
P A Foster

C J Fergusson 36
H E Ford 27
W R Gent 53 First Engineer
F A Graham 28 Cook
C A Burton
A C Hastings 30
A H Savidge 52
S E Scott 40
G W Short 34
C E Taylor 28
J Buck Lord Thomson's attendant
Samuel Church

WHO LOST THEIR LIVES IN THE AIRSHIP R 101 IN FRANCE
Sunday 5th October 1930

IN THE MIDST OF LIFE WE ARE IN DEATH

By kind permission of H.M. the King, the lying in State will be in Westminster Hall

A loss so great, a shock severe
To part with those we loved so dear,
Though great the loss, we'll not complain
But trust in Christ to meet again.

Printed by S. Burgess 8 York Place Strand, W.C.2

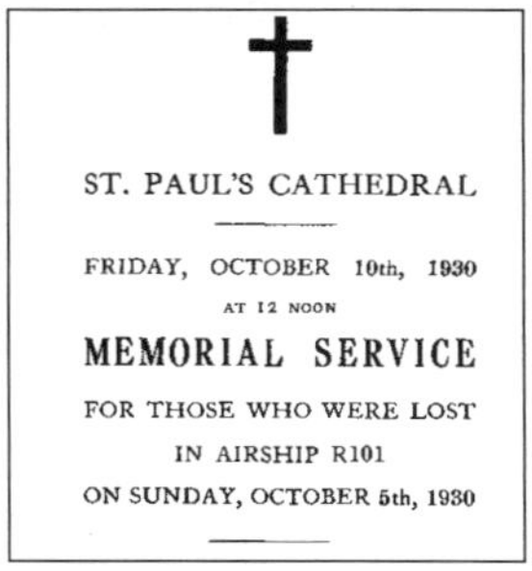
ST. PAUL'S CATHEDRAL

FRIDAY, OCTOBER 10th, 1930
AT 12 NOON

MEMORIAL SERVICE
FOR THOSE WHO WERE LOST
IN AIRSHIP R101
ON SUNDAY, OCTOBER 5th, 1930

CONCLUSION

The voyage by the *R-100* to Canada put the town of Saint-Hubert, now a borough in the City of Longueuil, on the map. More than 76 years after that event, people still talk about it. The success of the *R-100* and the subsequent failure of its cousin, the wrecked airship *R-101*c, marked an important turning point in the history of British aviation. The British government however, did not immediately give up building airships: on May 14, 1931, the British Prime Minister authorized a £400,000 project to upgrade the *R-100* as well as to build the R-102 (length of 250 metres and seven reversible Tornado engines). But in the month of December, he abandoned his political goal of uniting the colonies by utilizing such a means of transportation. The *R-100* was sent to the junkyard. It can be said therefore, that the destruction of the *R-101* led to the eventual downfall of the *R-100*. Considering that the overall program, including the construction of the mooring towers, had cost the British taxpayers close to £2,400,000, the government ultimately concluded that building airships was too ambitious. In the end, a wise decision.

On May 6, 1937, the *Hindenburg* exploded in the air while approaching the mooring tower at Lakehurst, New Jersey; 36 died. The horror caused by the images of the airship in flames put an end to the construction of lighter than air crafts; the *LZ-130 Graf Zeppelin* II, already built, had only made a few test flights before being stored and finally destroyed. Yet another page of mourning, one too many, was added to the black history of the airships. Moreover, the economic crisis was affecting England. It was not very reassuring to see huge sums of money invested in airship construction programs. From a purely technical point of view, airplanes by now could fly much faster than airships. The fundamental error was likely made when the expertise was split into two camps that failed to communicate with one another. If the same budget had been directed to the construction of only one dirigible, the quality of the final product would have surely been superior and, above all, political pressure, favouring one firm over the other, at any cost, would not have played a role.

Those airships could have flown for some thirty years inspiring admiration, curiosity, ambition. Even today, when a non-rigid dirigible flies over our heads or makes an appearance over a stadium during a football or a baseball match, it attracts great admiration among the spectators.

Never before had the Montérégie region of Quebec attracted so many tourists for an event of such short duration. One had to wait until Expo 67 to again see such numbers. The efforts of the Canadian government in this dossier contributed to the economic development of the region of Montreal. By November, 1927, the Saint-Hubert Airport had become an airmail postal depot. Later, it was established as a significant military airport. Even today, the airport zone represents a strategic investment for the new amalgamated City of Longueuil.

There is almost nothing left of the *R-100* and the *R-101*. Bolduc's song can still be heard, and the photos of that epoch are still available. The Saint-Hubert tower was demolished in 1938; the one in Cardington in 1943; in order to sustain the war effort, the metal had to be recycled. As a fire had destroyed the *R-101*, there are few souvenirs left. A few inhabitants of the regions of Allone and Beauvais, in France, have kept some pieces of metal from its frame. The *R-100* was dismantled; some heritage items such as pieces of its fabric, or the two wheels of the rudder, are preserved in museums. The Motel *R-100*, on Chambly Road, near the Saint-Hubert air base, has been around for many years. One street, located near the site of the mooring tower, bears the name *R-100*. Only those 80 years of age or older can tell us that they saw that marvel. When they speak of it, they do so with emotion. We believe the responsibility to safeguard this moment for our collective memory fell to us. We hope you share it.

The Saint-Hubert mooring tower, the last real souvenir of the *R-100*'s visit, is destroyed in the beginning of the month of January, 1938 and sold for scrap metal.
Louis Emard Collection.

Glossary of the R-100 and R-101

Allonne. Small village in France near the site where the *R-101* crashed.

Beauvais. Town in France where the funeral service of those killed in the catastrophe of *R-101* took place.

Booth, Ralph Sleigh. Captain of the *R-100*, he piloted the first test flight of the *R-100.*

Brancker, Sefton William (1877-1930). Director of British Civil Aviation, he was also an air vice-marshal.

Burney, Dennistoun Sir (1889-1968). General Manager of the Airship Guarantee Company, a subsidiary of the Vickers Company, for the manufacture of *R-100.*

<

Cardington. Site of the factory, near Bedford, of the *R-101* and place of departure for the *R-100* and the *R-101.*

Colmore, Reginald Blaney Bulteel (1883-1930). Lieutenant-Colonel for Aviation and Director of the Technical Development of the airships in England, he took part in the trip to Saint-Hubert with the *R-100.* He died in the crash of the *R-101.*

Howell. Building site in Yorkshire, of the *R-100.*

Irwin, Carmichael « Bird » H. (? -1930). Captain of the *R-101.*

Johnston, Ernest (1891-1930).
He was employed, in 1924, by Royal Airship Works. He also worked for Imperial Airways. Navigator for the *R-100* as well as the *R-101.* He died in the accident of the *R-101.*

Meager, George. First officer of the *R-100*, he was reputed to be a very meticulous person. Author of My Airship Flights, he was a Captain with the Royal Naval Air Service.

Richmond, Vincent Crane (1893-1930). Lieutenant-Colonel, he was hired in the Royal Airship Works, in 1924, as the designer of *R-101*. Very respected in his field, he was named Assistant-Director of the Technical Development of the airships.

Scott, George Herbert (1888-1930). Crew member in the voyage of the *R-100*, he was captain for the R-9, the first rigid airship to fly. He was a captain of the *R-34* at the time of the crossing, for the first time, of the Atlantic in July 1919. Major in the Royal Air Force, he conceived the British system of mooring of the airships. He supervised many tests on *R-100* and *R-101*.

Shute Norway, Nevil (1899-1961). Calculator-in-Chief for the team of R-100, well known novelist. In his autobiography can be found some reflexions on the construction of the *R-100* and the *R-101*. His book, excellent for certain details, is considered by some analysts as being quite biased.

<

Thomson of Cardington, Lord Christopher Birdwood (1875-1930). Secretary of State for Air in the Labour Government of James Ramsay MacDonald, he is held responsible for the decision to make the fatal trip to Karachi in spite of the many recommendations that the *R-101* was not ready.

Vickers. British company responsible for the construction of the airship *R-100*.

Wallis, Barnes Neville Sir (1887-1979). A doctor's son, Chief Engineer for the *R-100*, and employee of Vickers, he was regarded as a very bright yet solitary engineer. He also designed the *R-80*, the Wellington bomber of the Second World War and the "bouncing bomb" able to destroy hydraulic dams. In 1928, he accepted the silver medal from the Royal Aeronautical Society for his contribution to the development of the airships, and was named Sir in 1968.

Comparison of Airships

	R-100b	*R-101c*	Graf Zeppelin	Hindenburg
Length (m)	211,8	236,8	236,8	245,1
Diametre (m)	40,5	40,2	30,5	41,1
Ratio length-diameter	5,21 :1	5,9 : 1	7,7 : 1	5,9 : 1
Max speed (km/h)	130,4	100	128	135,8
Gas volume (m^3)	146 000	156 000	105 000	211 898
Engine brand	Rolls Royce Condor IIIB	Beardmore Tornado II C.1	Maybach	Daimler-Benz
Number of engines	6	5	5	4
Power per engine (h.p.)	650	525	530	1050
Fuel	Gazoline	Diesel	« Blau-gaz »	Diesel
Cylinders	12	8	12	16
Weight per engine (kg)	1 778	3 892	n.d.	n.d.
Gasbags	15	17	17	16
Distance covered (km)	23 100	8 575	1 600 00	n.d.
Hours of flight	276	127	17 177	3 088

LIST OF THE DECEASED IN THE ACCIDENT OF THE *R-101*

*** were on the trip of the R-100 to Canada**

ATHERSTONE, Noël Grabowski	First Officer
ATKINS*, George K.	Wireless Operator
BISHOP, Percy	Chief Inspector
BLAKE Richard	Engineer
BRANCKER, Sir Sefton	Director of Civil Aviation
BUCK, James	Valet to Lord Thomson
BURTON, Charles A.	Engineer
BUSHFIELD, Alexander	Aeronautical Inspection Directorate
CHURCH, Samuel	Rigger
COLMORE*, Reginald Blaney Bulteel	Director of Airship Development
CURRAN, John F. (HODNETT*)	Assistant steward
ELLIOTT, Frank	Wireless Operator
FERGUSSON, Christopher, J.	Engineer
FORD, Hector E.	Rigger
FOSTER, Percy A.	Rigger
GENT, William R.	First Engineer
GIBLETT*, Maurice A.	Chief Meteorological Officer
GRAHAM, Eric A.	Steward
HASTINGS, Alfred C.	Engineer
HUNT, George W.	Chief Coxswain
IRWIN, Carmichael H.	Captain of the R-*101*
JOHNSTON*, Ernest L.	Navigator
KEELEY, Spencer T.	Chief Wireless Operator
KEY, Thomas A.	Chargehand Engineer
KING, William H.	Engineer
LITTLEKIT, Maurice F.	Engineer
MASON, Christopher H.	Assistant Coxswain
MEGGINSON*, Thomas W.	Steward
MOULE, Wilfred	Engineer
NORCOTT, Arthur W.J.	Rigger
O'NEILL, William H.L.	Deputy Director of Civil Aviation, India
OUGHTON, Leonard F.	Assistant Coxswain
PALSTRA, William	Australian Government representative
POTTER, Walter A.	Assistant Coxswain
RADCLIFFE, Walter G.	Rigger
RAMPTON, Martin G.	Rigger
RICHARDSON, Arthur J.	Rigger
RICHMOND, Vincent C.	Assistant Director Technical
ROPE, Frederick M.	Squadron Leader
RUDD, Ernest G.	Rigger
SAVIDGE*, Albert H.	Chief Steward
SCOTT*, George H.	Assistant Director Flying
SCOTT, S.E.	Chargehand Engineer
SHORT, G.W.	Chargehand Engineer
STEFF*, Maurice H.	Second Officier
TAYLOR, Cecil E.	Rigger
THOMSON, Lord	British Air Minister
WATKINS, Albert H.	Engineer

List of the Survivors of the *R-101*

BELL, A.V.	Engineer
BINKS, T.H.	Engineer
COOK, Alf J.	Engineer
DISLEY*, A.	Electrician
LEECH, Mr A.H.	Foreman Engineer
SAVORY, V.	Engineer

List of Tragedies Involving Rigid Airships (1921-1937)

Dixmude	December 21, 1923	Mediterranean, near Sicily	52 dead
Shenandoah	September 3, 1925	United States, Ohio	14 dead
R-101	October 5, 1930	France, near Beauvais	48 dead
ZR4. Akron	April 4, 1933	Atlantic, near East coast of USA	73 dead
ZR5. Macon	February 12, 1935	United States, California	2 dead
Hindenburg	May 6, 1937	United States, Lakehurst, N. J.	36 dead

List of the Three Airships Which Had Crossed The Atlantic, Non-stop, Before the R-100

1919 The British airship commanded by captain Scott
1924 The German airship *Zeppelin ZP3*, later renamed the Los Angeles by the Americans
1928 The German airship *Graf Zeppelin*, commanded by Hugo Eckener

Annotated bibliography

ABBOTT, Patrick and Nick WALMSLEY. *British airships in pictures. An illustrated history 1784-1998*, House of Lochar, 1998, 112 p.

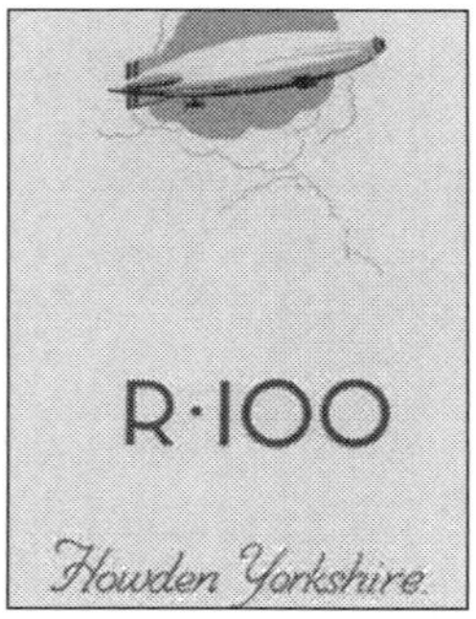

AIRSHIP GUARANTEE COMPANY. *Airship R.100*, 1929, 23 p.
Small booklet explaining in a simple way what an airship is, its advantages, but also its disadvantages. Astonishing promotional booklet by the company which built the airship.

AJALBERT, Jean. *Le R-101 sur Beauvais, route des Indes*, Les Éditions des Portiques, 1931, 77 p.
Very beautiful writing. It is the first book on the subject published in France.

BIBESCO (Princesse). *Le destin de Lord Thomson of Cardington*, Flammarion, 1932, 278 p. The princess tells about her adventures with Lord Thomson and slips in a few personal details.

BURNEY, sir Charles Dennistoun. *The World, the Air and the Future*, Alfred A. Knopf, 1929, 356 p. A book which made many waves at the time of its publication. Burney presents us with a study including important reflexions. He questions the mooring towers and the viability of the airships even if he himself acted as manufacturer of the *R-100*. Very rare book.

CANADIAN GEOGRAPHICAL JOURNAL. *The story of the R-100*, Vol.1 nº 6, oct. 1930.

CAVE-BROWN-CAVE, T.R. « *The machinery installation of airship R.101* », Institution of automobile engineers, 1928-1929, vol. XXIII, p. 99-123.
A study on the construction of the *R-101* and particularly on the functioning of its engines.

CHAMBERLAIN, Geoffrey. *Airships-Cardington, A history of Cardington Airship Station and Its Role in World Airship Development*, Terence Dalton, 1984, 223 p.
A relevant history of the Cardington base, including an important section on the history of the airships *R-100* and *R-101*.

COCHRAN, E.B. *The British dirigible program 1919-1930*. Self-published study, in the form of a spiral notebook, 1980.

COLLIER, Basil. *The Airship : A History*. New York, G.P. Putnam's son, 1974, 271 p.

COUNTRYMAN, Barry. *R100 in Canada*. The Boston Mills Press, 1982, 128 p.
The most complete study of the *R-100*. Excellent book, but hard to find.

DIRIGIBLE. *R100 and R101,* 1999.
A selection of articles published by the bulletin of the Airship Heritage Trust, *Dirigible*.

DUBOIS, Jean-Yves. *Il était une fois... le R-100.* Société d'histoire de Saint-Hubert, 1992, 76 p.

FLIGHT.
An indispensable review, as much for the quality of its photos as for the aeronautical news articles of that epoch.

FORTIER, Rénald. *Le R.100 au Canada*. (Ressource électronique). Musée national de l'aviation, 1999, 24 p. The author had published, in 1998, an exhaustive study: *The R-100 has come: Canada and the imperial program of transportation by airships 1924-1931*. 176 p.

JOHNSTON, E. A. *Airship navigator. One man's part in the British airship tragedy 1916-1930*. Skyline Publishing, 1994, 205 p.
Excellent account of the voyages of the airships *R-100* and *R-101*. The author is the son of Ernest L. Johnston, who participated in the voyages of the *R-100* and the *R-101*.

LA NATURE (revue française).

LEASOR, James. *The Millionth chance, The story of the R.101*, Reaynal and Company, 1957, 244 p.
A well-written and well-documented story.

MASEFIELD, Sir Peter. *To ride the storm*, Londres, William Kimber, 1982, 560 p.
The biggest classic in the history of *R-101* and the best-documented history. However, the author seems biased in favor of Lord Thomson and the quality of *R-101*.

MEAGER, George. *My Airship Flights 1915-1930*. William Kimber, 1970, 239 p.
The author made the trip aboard the *R-100* as an officer. He relates in detail his experience.

CANADIAN FOREIGN AFFAIRS DEPARTMENT
Documents Related to Canada's Foreign Relations, Volume IV, 1926-1930.

MORPURGO, J.E. *Barnes Wallis : A Biography*, Londres, 1972, 400 p.
A very good biography. Unfortunately the part about to the development of the *R-100* leaves us wanting more.

SHUTE, Nevil. *Slide Rule, an autobiography*, House of Stratus, 2000 (édition originale 1954), 176 p.
A very controversial vision of the competition between the *R-100* and the *R-101*.

THE ADVERTISING CLUB OF MONTREAL. *Visit of His Majesty's Airship R.100 Canada – 1930*, 96 p.
Souvenir program. A real collector's item.

THE STATIONERY OFFICE. *R.101: The Airship Disaster, 1930*, London: 1999, 162 p.
British investigative report on the tragedy of the *R-101*.

VILLEROY, Louis Gérard. *Il était une fois un Géant de l'air : le R-101*, Historical and folkloric booklets of Oise J.P. Sohyer publisher, 1972, 52 p.
Very well-done. Very rare.

LE NEVE WALMSLEY, Nick. *R101. A pictorial history*. Sutton Publishing, 2000.
The author presents a part of the collection of photographs from the Airship Heritage Trust. It is the most beautiful book concerning the *R-101*.

WATTEBLED, Laurent. *La catastrophe du R-101*, 1990, 180 p.
Collection of quotations from newspapers and Villeroy's book. It is not regarded as a true study. On the other hand, the book is extremely well illustrated and it stresses the accident and the participation of the people of Beauvais and Allonne. Interesting tool for any researcher, since it offers a rigorous scan of the local newspapers.

Daily newspapers: *Chicago Daily Tribune, Daily Express* (Great Britain), *Daily Mirror* (Great Britain), *La Presse* (Montreal), *Le Devoir* (Montreal), *The New York Times, San Francisco Chronicle, The Gazette* (Montreal), *The Montreal Daily Star.*

APPENDIX
OFFICIAL REPORT OF THE FRENCH NATIONAL POLICE (GENDARMERIE)

SECOND LEGION. Company of Oise, n° 879/2

DISASTER

Destruction by explosion and fire of the English airship *R-101*.

REPORT

By Squadron Leader Serin, Commander of the Gendarmerie of Oise.

On the destruction by explosion and fire of the English airship *R-101* and almostall of its crew and passengers.

I. - THE EVENT

On Sunday October 5, around 2 a.m., the partly illuminated English airship *R-101*,whose trip to India had been announced by the newspapers, was flying over Beauvais with all engines roaring and trailing traces of flames from its exhaust ports.

It was an exceptionally dark night. Rain fell in gusts in a rather violent westerly wind.

The *R-101* was flying very low, so low that many observers awaiting the passage of the balloon, or who had been awaken by the noise of the engines, had the clear impression that something abnormal was happening on board. According to some, it seemed that "the engine housings skimmed the chimneys".

Soon, however, the airship moved away from Beauvais in a south-eastern direction, remaining nevertheless visible because of its lights. But, abruptly, they all went out (except for one or two according to witnesses) and, suddenly a gigantic hue lit up the sky just as a very loud explosion was heard, immediately followed by such a rush of air that some believed in earnest that a powder magazine or an ammunition dump, or the Beauvais gasworks, had exploded. It was, at that moment, exactly 10 minutes past two in the morning.

II. - ALARM

This explosion, by itself, had set off the alarm. From all over, good-willed people moved into action; the least aware, after having opened their windows, quickly understood that the intense hue which lit up the buildings of the city and the countryside, announced a significant disaster. Which is why, in less than 20 minutes, by 2:30, rescuers were on the site of the catastrophe, located in the commune of Allonne, in the place called "Les Coutumes", on the edge of the Fecq Woods, five kilometres to the south-east of Beauvais and halfway between National Road no 1 and Major Road n° 135.

The first on the scene, obviously, were the inhabitants of the nearby villages: Allonne, Warluis and Frocourt, then the Beauvais observers of the airship who, having witnessed the passage of the *R-101*, seen the immense hue, and heard the explosion, had deduced the tragedy.

III. - RESCUE

The first rescuers on the scene discovered:

The unscathed survivors, just shaken up. Crewmembers, seriously or slightly wounded. A total of eight people were directed at once to the hospital of Beauvais or a private clinic of the city to be hospitalized there, or to be bandaged, or simply to be comforted. Other rescuers trying to approach the balloon, noted that the wreckage lay in a North- West - South-East direction, its prow turned towards Paris; the front and the centre of the balloon being completely crushed on the ground with no form but one indescribable tangle of steel wires and twisted beams. Only the frame of the rear portion was apparently intact, but, of course, long stripped of its fabric by the flames.

Two thirds of the balloon (the front and the centre) however, were still in flames, because, once the fabric of the superstructure got torn and burnt by the explosion, the engine housings, the cabins and the machinery and engines rooms, built out of light wood or separated by fabric, continued to feed the blazing inferno, rekindled in some places by fuel leaking from large tanks spread out almost everywhere.

These tanks, whose contents were not precisely known, but thought to be full of fuel, constituted, in everyone's opinion, a permanent danger for the rescuers. According to those directing the rescue, it appeared that a second catastrophe could occur if these tanks exploded spreading ignited liquids over the people around the blazing inferno or who were preparing to enter the carcass. Had they not heard, indeed, two or three explosions, undoubtedly lesser in magnitude than the first, accompany the first explosion? They were, it was said later, small fuel tanks intended for the small engines used to start the balloon's diesels; but, until the break of dawn and in particular until the formal declaration of engineer Leech, survivor of the catastrophe, the large tanks of the airship were thought to contain gasoline (whereas they were simply filled with fuel oil). The most elementary prudence tied to the desire that each had to do his duty, required the restraint of the ardor of all those burning to be of service.

Furthermore, the *R-101* was in flames at 10 minutes past two. Under its carcass, tangled up in thousands of beams and wires, lay, at the moment when the first rescuers arrived (approximately 20 minutes later) those who would not make it out alive from the airship. The intense heat generated from the aircraft stopped any attempt made to enter it. A funereal silence hovered about the balloon from which no complaint, no cry escaped. All that was heard was the crackling of the fire and the deeply disturbing noise from the carbonizing of the bodies.

IV. - THE RECOVERY OF THE BODIES

Engineer Leech, a survivor, despite burned hands and a head wound, had returned to the site and had specified that at the time of the catastrophe, passengers and crewmembers were in bed, except for the twelve crewmen operating the craft from their respective posts. He advised that all the bodies were grouped, mostly in the front of the airship, and, a smaller group, in the centre. Among those missing, he cited the names of Lord Thomson, British Air Minister, and Sir Sefton Brancker, Under-Secretary of State for War.

The most ardent amongst the rescuers entered the frame despite the extremely hotbeams and, walking on still burning member, started to remove some of the bodies. The rescuers burned their hands while raising them because, horrible as it may be to describe, the flesh of these bodies was cooked, carbonized and had to be removed, including their intestines, from thousands of tangled steel wires engulfing most of the corpses.

Around seven o'clock, approximately fifteen bodies had been removed and decently arranged along a thicket, each one of them wrapped or covered with a white cloth. As soon as the first coffins arrived, the corpses were placed in them by gendarmes and firemen, following a first attempt at identification resulting from interesting discoveries made around the corpses or on the bodies. The objects (rings, watches, cigarette cases, etc.) were placed in a numbered envelope in the coffin.

At 9 o'clock close to 30 bodies had been recovered.

Also the time of the arrival of the French Air Minister, greeted near the airship by the Prefect of Oise.

V. - THE INVESTIGATION

The technical investigation was ordered, at the same time, by the French Air Minister and by General Salmon who arrived at the site by plane, as the delegate of the British Government.

VI. - IDENTIFICATION AND FUNERALS

A renewed attempt at identification took place, at six in the morning, in Allonne. As a result, five more bodies were identified.

Once identified, the coffins with the deceased were transported to the town hall of Beauvais where they were received by an honour guard.

This long and painful identification procedure made it possible to establish that, by comparison with the list sent from Bedford by plane, the total determined by adding the eight survivors and the 46 bodies discovered in the debris coincided precisely with the number of departing passengers, namely 54. The theory of a stowaway (based on an incinerated shoe, of small size, supposedly belonging to a woman), was therefore abandoned.

Spirits were calmed once it was definitively established that no one was missing, whether lying wounded in the countryside, or trapped under the remains of the airship.

On the other hand, it was certain that Lord Thomson, Sir Sefton Brancker, Commander-in -Chief Irwin and his second in command, Lieutenant Scott, were among the dead.

TEXT BY A NUN FROM THE MONASTERY NOTRE-DAME DE CHARITÉ DU BON-PASTEUR D'ANGERS CONCERNING THE VISIT OF THE R-100 TO CANADA.

While God thus multiplied his gifts to his children, time also moved along quickly. We would not like to ignore the soothing distractions that procured for us, after its departure, the arrival of the famous *R-100* airship. That grandiose demonstration engendered an extraordinary aerial meandering for our country. The mooring mast being located near our gardens, we watched, in space, the splendid evolutions which attracted innumerable spectators. The gigantic airship flew above our convent before receiving the benevolent reception of the tower as if extending its arms at sunrise. Ideal spectacle inciting meditation as well as admiration. It showed us God acting on human thought where He creates wonders the world, not thinking about the Author, enjoys attributing to the genius of the scientists.

After having interested, for fifteen days, an entire people moved by contemplating it, the *R-100* resumed its airborne course towards Cardington. What will happen next, everyone? Our good Praying Mother entrusted it to Mary, guide of travellers; perhaps it owes to herfervent prayer preservation from the the sad fate of its brother the *R-101*...

N.B. This text was written in 1940.

SAINT-HUBERT AIRPORT STAFF 1928-1930

Senior staff

Geo. G. Wakeman,	inspector Eastern Airways
S. Graham,	inspector
Major R. Dodds,	inspector

Staff

Name	Post	Salary per month $	Hired on
Office staff			
Mlle F. Beasant	Stenographer	90	January 15, 1930
Mlle A. Knight	Stenographer	90	May 19, 1930
Mlle H. A. Macaulay	Stenographer	90	July 16, 1930
B. Yellion	Clerk	130	May 13, 1930
Mlle A. Boissy	Operator	90	July 19, 1930
R. Galaise	Operator	60	January 7, 1930
M. Daoust	Night operator	90	June 7, 1930
J.A. Fallon	Clerk	90	April 10, 1930
R. Caumartini	Janitor	90	April, 1930
G. Prindle	Janitor	90	1929
Garage			
F. Vincent	Hangar foreman	125	December 15, 1929
F. Hebert	Driver	125	September 20, 1928
H. Lamontagne	Worker	90	June 9, 1928
Pumps			
C. St Laurent	Foreman	135	June 1, 1929
T. Hebert	Worker	90	April 1, 1930
L. Thuot	Worker	90	April 1,1930
Hangar n° 2			
A. Flannigan	Hangar foreman	130	April 1, 1930
N. P. Prindle	Day laborer	90	August 1, 1929
O. Monette	Day laborer	90	April 13, 1930
J. Carr	Day laborer	90	February 21, 1930
W. Bruce	Day laborer	90	April 2, 1930
A. Schmidt	Day laborer	90	April 1, 1930
L. Schmidt	Day laborer	90	April 1, 1930
R. Mongeau	Night watchman	75	April 1, 1930
O. Monette	Night watchman	90	
Weather center			
W. F. Carmichael	Weather analyst	145	July 16, 1928
H. Newton	Teletype operator	110	July 2, 1930
C. Fontaine	Clerk	90	July 2, 1930

Control tower staff

W. L. Laurie	Radio officer
W. G. Megill	Assistant to the officer at mooring tower
W. C. Christensen	Control chamber
D. W. Mitchell	Control chamber
C.S.M. G. C. Gould	Radio operator (control chamber)

S. D. Ranns	Radio operator (control chamber)		
C. V. Matthews	Radio operator (control chamber)		
D. McKinnon	Radio operator (control chamber)		
J. C. Cavana	Radio operator (control chamber)		
H. L. D. MoCormick	Radio operator (control chamber)		
C. W. King Mechanic,	Radio Beacon, La Prairie		
Mooring tower			
A. R. Pressey	Officer in charge		
C. DeYoung	Superintendant		
Parker C.P.O.	Foreman	200	April 1, 1930
Williams C.	Team leader	200	April 1, 1930
H. Kinch C.	Team leader	200	April 1, 1930
MacDonald	Mechanic,	150	April 1 1930
Dowson	Mechanic,	150	May 14, 1930
Boisvert	Elevator	100	April 1, 1930
Beaulac	Worker 100		April 1, 1930
Williams	Rigger	100	May 2, 1930
Matthews	Rigger	100	April 30, 1930
Grafton	Rigger	100	April 1, 1930
Grafton	Rigger	100	May 8, 1930
Knight	Rigger	100	May 12, 1930
Knoakes	Rigger	100	May 12, 1930
Eccles	Rigger	100	June 25, 1930
Charbonneau	Supervisor	100	April 1, 1930
Hydrogen plant			
Alley	Mechanic	£12 per week	
H. Temple	Team leader	175	April 1, 1930
Irwin	Team leader	110	May 23, 1930
Mechanics			
S. M. G. F. Millward	Mechanic		
Lyon	Foreman	200	
Workers under contract			
Charles Martin	Foreman 90¢ / hour		April 1930
Saint-Germain	Carpenter 85¢ / hour		April 1930
Fournier	Carpenter 85¢ / hour		April 1930
S. Mainville	Worker 40¢ / hour		April 1930
Savaria	Worker 40¢ / hour		April 1930
Laberge	Tractor 75¢ / hour		April 1930
Pitre	Worker 40¢ / hour		June 27, 1930

PCommemorative plaque for the *R-100*, erected near the Saint-Hubert Airport in 1990.

R-100 Street, near the location of the mooring tower.

R - 100
MOTEL
T.V. AIR CONDITIONED
NO VACANCY
MOTEL R-100
4335 CHEMIN CHAMBLY
5 Minutes de Montréal
ST.HUBERT

Postcard.

From the same author

Italo Balbo's Transatlantic Flight (1933): 24 Italian seaplanes in America

The cruise was a mass transatlantic flight from Orbetello, Italy, to the Century of Progress International Exposition, Chicago, Illinois. The expedition, organized by the Italian Regia Aeronautica, began on July 1, 1933 and ended on August 12 of the same year. It comprised 25 Savoia-Marchetti S.55X seaplanes crossing the Atlantic Ocean in formation, according to the press of the time the greatest mass flight in aviation history. The Italian Squadrons, led by General Italo Balbo, were welcomed enthusiastically in the Netherlands, the UK, Iceland, Canada and particularly in the United States of America, where they became known as the Italian Air Armada. A publicity success for Fascist Italy, Balbo further viewed the expedition as a pioneering step towards commercial flights across the Atlantic.

In Chicago, a crowd of over 100,000 gathered to cheer the arrival of Balbo's fleet on the evening of July 15. July 15 was proclaimed Italo Balbo Day and on July 17 Chicago's Seventh Street was renamed Balbo Avenue. Balbo received a gold key to the city from Mayor Kelly. United States Commissioner to the Century of Progress Exposition Harry Stewart New said that if Balbo remained in the US he might be elected President. In a ceremony that Balbo found strange, they initiated him into the Sioux tribe as "Chief Flying Eagle". Being very cautious about his public image, Balbo was unsure whether to accept and agreed only after hearing that former US President Calvin Coolidge also received such an honour. The Italians were "hailed by millions" on their arrival in New York City according to the front page of the *New York Times* of July 20. 1500 persons watched the planes from the Empire State Building. Balbo was quoted saying that "New York's welcome exceeded all expectations, more so even than Chicago's".

On July 20 Roosevelt invited Balbo to lunch at the White House, where he went with a select group of his most senior aviators. The President offered his congratulations and tried to persuade Balbo to prolong his stay in the US to start a countrywide tour. Balbo enjoyed the lunch with Roosevelt, which he found to be "very cordial and direct in his manners, like all Americans".

A crowd of 60,000 attended the celebrations at Madison Square Garden. The city of New York organized a parade down Broadway on motorcycle escort on July 21 which made a great impression on Balbo, who said "in our lives, it would be extraordinary to experiment something as grandiose again.

Source : *Wikipedia*

Available on Amazon

Made in the USA
San Bernardino, CA
29 January 2020

63772665R00113